Exterior Ballistics
of Small Arms

Exterior Ballistics of Small Arms

Companion to Exterior Ballistics
with Applications

Gjergj Klimi

To order additional copies of this book, contact:
Xlibris Corporation
1-888-795-4274
www.Xlibris.com
Orders@Xlibris.com
55122

CONTENTS

Chapter 3 Computation Of The Firing Range

To my wonderful nieces
Viola and Nensi Klimi

PREFACE

The "Exterior Ballistics of Small Arms" is a manuscript mainly about the flight of projectiles of small arms and, at the same time, represents an extension to "Exterior Ballistics with Applications—Skydiving, parachute Fall, flying fragments", by Gjergj Klimi, already published by Xlibris in July 30th, 2008. In other words, the book can be seen as a companion to the "Exterior Ballistics with Applications". It contains the Exterior Ballistics PC programs that was not possible to be included in "Exterior Ballistics with Applications" as well as illustration examples and exercises that can be solved mainly using the Exterior Ballistics PC programs presented in this book.

The present book is addressed to amateurs and professionals interested in exterior ballistics, and in shooting with small arms, hunting and sporting rifles, and in general in the field of military and applied science.

The simple undergraduate mathematics that is used to present the material and the PC programs makes the book attractive to the amateurs and training experts that continuously practice to improve the accuracy of shooting with small arms. At the same time, the manuscript might serve as a basis to understand better the particular characteristics of the theory of the exterior ballistics of small arms.

The system of units used in the book is the SI system. For readers that are unfamiliar with the metric system it is not difficult to become accustomed and use the materials presented in the book to benefit from the simple illustrations, exercises and PC programs that, at the same time, give answers to many problems the shooters might encounter in practice.

For better understanding of the information the book contains, the reader should refer to the "Exterior Ballistics with Applications", although most of the material is clear enough to be assimilated without the necessity to go through it.

In the book, we use Siacci's methods and the Siacci's drag function to study the projectile trajectory. The Siacci's Drag function is abandoned nowadays by most of the ballisticians, but not the Siacci's methods that still are used in the ballistics of small arms.

To overcome the difficulties that the readers might encounter because they are mostly familiar with other than Siacci's drag function, in the book are given the appropriate Siacci's ballistics coefficients of some standard bullets and are shown the methods used to find the Siacci's ballistic coefficients.

The small firearm shooters find in the book simple theoretical explanations to some basics concepts and characteristics of exterior ballistics and practice of shooting with small arms such as the ballistics coefficient, rifleman's rule, inclined fire, mountain firing, firing in non-standard atmosphere and in presence of wind, etc. All of those explanations as well as the proofs of some fundamental rules of exterior ballistics are based on the flight of the projectile in presence of drag.

The long practice of modern exterior ballistics has shown that the ballistics coefficient of any bullet is a variable quantity, but to match the theoretical data with the experimental data of firing it is used a fixed ballistics coefficient valid only within a relatively small range of departure angles, or within an interval of projectile velocities. The property of a proper ballistics coefficient of a projectile that can be considered fixed for such intervals gives satisfying accuracy in estimating the elements of the projectile trajectory.

The ballistics coefficient and the drag function as well are fundamental factors necessary to study the projectile flight and to estimate the elements of the projectile trajectory. The appropriate determination of the ballistics coefficient of a projectile, and of a bullet particularly, is the key factor to describe the theoretical trajectory and to match it with a satisfying degree of accuracy with the real trajectory.

The determination of an "appropriate" Siacci's ballistics coefficient for a bullet is facilitated by the use of the PC programs that help the reader to find easily the Siacci's ballistics coefficient using his/her own experimental data, or the data given by bullet manufacturers or other sources.

The theoretical practicalities are illustrated and verified through examples that are related with the practice of shooting with small arms.

The book contains 76 examples and exercises, and 19 PC Programs.

To use the PC programs written in Qbasic the reader must have the Qbasic.exe file and then must copy each of the PC programs using Qbasic. exe to create electronic PC programs. Once the programs are in separate electronic files, it is easy to execute any program.

The PC programs Mortar.Bas and Sierra.Bas are not present in the book, since for the respective projectile trajectories we can use the general programs Rameco.Bas and Rcpoint.Bas.

We only need to know the Siacci ballistics coefficients of the projectile of a mortar, or the Siacci ballistics coefficient of a Sierra bullet. Some ballistics coefficients of the Sierra bullets are shown in section 1.1.

At the end of the book is shown the Errata for the book *Exterior Ballistics with Applications—Skydiving, Parachute Fall, Flying Fragments,* by Gjergj Klimi, Xlibris Corporation, 2008.

I apologize for not having the publishing possibility to attach a CD of electronic files of the PC programs with the manuscript.

Hereafter, we refer to the manuscript "Exterior Ballistics with Applications—Skydiving, parachute Fall, Flying fragments", by Gjergj Klimi, Xlibris, 2008 simply as the "Exterior Ballistics with Applications".

George Klimi, Ph.D.
January 2009

ACKNOWLEDGEMENTS

I would especially like to thank Mr. John C. Schaefer (USA) (http://www.frfrogspad.com) and Edoardo Mori, Esq. (Italy) (http://www.earmi.it/) for their encouragement in writing the book and for their valuable expertise in the theory of Exterior Ballistics and the practice of shooting with small arms communicated to me during the process of writing the book.

DISCLAIMER

N either the author nor Xlibris accepts any responsibility for any problem (errors, damages, injuries, etc) that might occur in practice of shooting applying the PC programs to solve practical ballistics problems.

Though the PC programs most of the time give good outcomes, it might occur that the reader obtains wrong solutions, for example as result of incorrect input of data, calculation of erroneous preliminary data needed to run a program, incorrect use of PC programs, use of the data given by different authors, use of data obtained experimentally, etc.

The user of the PC programs should be cautious and use the common sense to judge the results obtained using the PC programs.

1

EXTERIOR BALLISTICS OF SMALL ARMS

Introduction

In the practice of shooting, a marksman frequently uses the range table of an arm to control the fire in order to hit the target that is not necessary located at a perfect horizontal range described by the range table. The firing situations are not ideal, not only because of the intermediate ranges that are not present in the range tables, but as well because of the changes in the angle of site, in the altitude of shooting, as well as because of changes in the atmospheric conditions, etc.

Therefore, the exterior ballistics of small arms elaborates some useful methods to make firing practical, accurate and effective in a terrain of fast moving or hiding targets.

The sight arms are effective in the ascending part of the trajectory, and their fire is called "direct fire". The trajectories of projectiles fired by the sight arms, such as some navy and tank cannons, some infantry arms, hunting and sport rifles, etc., have some interesting characteristics that allow us to set the angle of sight or the departure angle for the horizontal and inclined shooting based on the horizontal range table of the respective firearm.

In this chapter, we study flight trajectories of the projectiles fired mainly by small arms to explore some characteristics of the projectiles flight and apply them in the practice of shooting.

Note that we use the term "angle of sight" to describe the angle that the line of bullet departure forms with the line of sight (LOS) (Figure 1 and Figure 2).

Recall that the characteristics of the standard atmosphere at the sea level have the following values:

Virtual temperature, 289.08 degree Kelvin (15 degree Celsius); air density, 1.205kg; pressure, 750mm Hg; the water vapor pressure at 15 degree, 750 mm Hg and 50% humidity is 6.35mm Hg; speed of sound 340.83m/s; no wind is present.

The values of the virtual temperature, the density and pressure at any altitude, when the atmosphere is standard, can be estimated using formulas (1.12.1)-(1.12.4), section 1.12.

The atmosphere is not standard if at least one of the characteristics of the atmospheric air at the sea level is different from the standard values.

For more information on the standard or non-standard atmosphere and the dependence on the temperature of the speed of sound, see chapter eight of the book "Exterior Ballistics with Applications".

In all PC programs that are used for computation of projectile trajectory in the standard atmosphere we have included as well the range wind and the cross wind. The elements of the projectile trajectory in the standard atmosphere are computed employing the PC programs considering zero the speed values of the range wind and the crosswind.

In Figure 1, we show a simple draw of a bullet trajectory zeroed at the point T, at the distance x_T from the gun, as well as the other elements related with the projectile trajectory. The gun and the target are at the sea level, but they can be in the same altitude over the sea level.

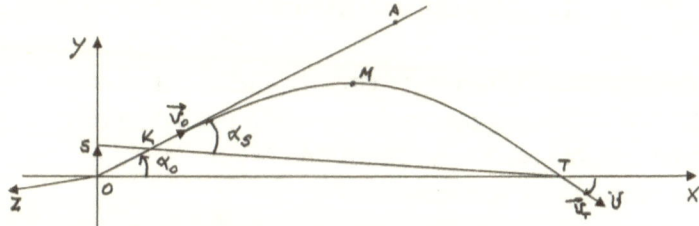

Figure 1—Projectile Trajectory

OMT—Projectile Trajectory; OT—Horizontal Range; OA—Line of Departure;

$\vec{v}_0$—Departure Velocity; TOA—Departure Angle; OS—Sight of the Gun;

ST—Line of Sight; TKA—Angle of Sight; M—Trajectory Vertex;

O—Origin of the Coordinates, T—Terminal Point; XTU—Terminal Angle

$\vec{v}_T$—Terminal Velocity.

In Figure 2, there are shown the elements of the trajectory of a bullet zeroed at the inclined range OE = D. The bullet is fired from a small arm.

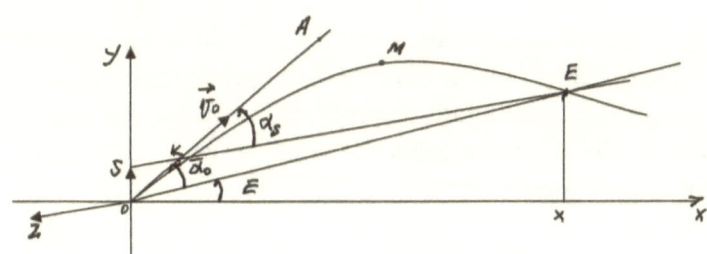

Figure 2—Inclined Shooting

OME—Projectile Trajectory; OE—Inclined Range; OA—Line of Departure;

$\vec{v}_0$—Departure Velocity; EOA—Super Elevation Angle; EOA—Angle of Sight;

S—Sight of the Gun; ST—Line of Sight; EKA—Angle of Sight; M—Trajectory Vertex;

O—Origin of the Coordinates, and the Location of the Muzzle of the GUN.

P—Point on the Trajectory, E—Terminal Point; XEU—Terminal Angle

$\vec{v}_T$—Terminal Velocity.

1.1 Form Factor and the Ballistics Coefficient of Bullets

The form factor (form coefficient) and the ballistics coefficient (BC) of a bullet are important parameters in exterior ballistics that allow us to obtain acceptable results for the trajectory of the bullet flight and to construct the ballistics and range tables.

In my already published book "Exterior Ballistics with Applications", p.345-347, Xlibris, 2008, we consider the ballistics coefficient as a parameter that is used to match the results obtained theoretically for the projectile trajectory with the results of experiments. We have seen two models as appropriate for the ballistics coefficient:

- The model that considers the ballistics coefficient a function of projectile speed;
- The model that considers the ballistics coefficient a function of the departure angle.

Nevertheless, in the practice of Exterior Ballistics, for a given projectile it is given a form coefficient (or a corresponding ballistics coefficient) that has a constant value for a small interval of departure angles or a range of projectile speeds. An appropriate constant ballistics coefficient gives approximate results especially for the trajectories of bullets within the effective range of small arms (usually till 600-800 meters for the arms with non-optical sights).

For large ranges, or for different departure angles, the use of a constant form coefficient estimated for relatively small ranges gives results that usually are far from the experimental results. In other words, for relatively long ranges (600-1000 meters or over), especially for the modern small arms and cartridges, and for the sniper arms, the accuracy of shooting is not satisfactory if the ballistics coefficient of bullet is measured for small horizontal ranges (usually around 100 meters), ref. section 1.3, "Determination of an Appropriate Ballistics Coefficient".

Anyway, as we will see, we can use some constant ballistics coefficients measured for different horizontal ranges: 500 meters, 800 meters 1000meters, etc.

For accurate results, especially for sniper shooting, it is necessary to use a ballistics coefficient that is a function of the projectile speed, or a function of the angle of departure, or to use more than one ballistics coefficients for different firing ranges, or departure angles.

The ballistics coefficient "c" and the form coefficient "i" of a bullet fired from infantry rifles, submachine guns, hunting rifles, etc. are related by the equation:

$$c = 1000 \frac{i \cdot d^2}{m} \cdot$$

(1.1)

Usually, some manufacturers of cartridges, for a given bullet give a form coefficient or a ballistics coefficient that corresponds to a non-Siacci function of resistance, for example to G1, G_2, G_7, or other drag functions. To use the PC programs presented in this book there are given the Siacci ballistics coefficients of some bullets.

In this section, we will see again some procedures that allow us to find the Siacci form factor and the ballistics coefficient of a projectile when we know a non-Siacci form factor or the ballistics coefficient of the same projectile.

The determination of an "appropriate" Siacci's ballistics coefficient for a bullet is facilitated by the PC programs that help the reader to find easily the Siacci ballistics coefficient using his/her own experimental data, or the data given by bullet manufacturers or other sources.

The following example illustrates the procedure we use to estimate a Siacci form coefficient when we know a non-Siacci form coefficient, while example 2 illustrates the opposite problem, i.e. illustrates how to estimate a non-Siacci form coefficient and corresponding BC when it is known the Siacci form coefficient, or the Siacci BC. The ballistics coefficients estimated in this way are approximate. They should be verified and adjusted using the PC programs coeff.bas, Anglec.Bas, etc, as well as the data of bullet manufacturers, or other sources.

The best way to find a proper ballistics coefficient is elaboration of the ballistics data obtained in the practice of shooting using the PC programs presented hereafter.

Example 1.1 Siacci Ballistics Coefficient

Estimate an average form factor for the bullet, caliber 0.30 Ball M2 (considering the following data (ref. McCoy R.L., "Modern Exterior Ballistics", p.113, Schiffer Publishing ltd, 1999) :

- Form coefficient that corresponds to G_8 Drag Function is 1.076.
- Speed Interval: 853.44m/s-472.44m/s.

We assume that the form coefficient of the bullet is determined as the average value of the form coefficients in the given interval of speeds.

Solution

The Siacci form coefficient of the given bullet can be estimated using the following formula (see G. Klimi, "Exterior Ballistics with Applications", p.94, Xlibris, 2008):

$$ i = 0.0014196 \frac{v^2}{v - 240} i_n C_n (\frac{v}{a}) \tag{1.2} $$

where "v" is the projectile speed, and "a" is the speed of sound in air on the ground level, (a = 340.83m/s). The ratio (v/a) is the Mach number.

To find an average form coefficient we need to find an average Mach number and the corresponding average bullet speed.

The interval of speeds, 853.44m/s-472.44m/s, corresponds approximately to the following interval of Mach numbers, 2.5-1.39. Since the drag coefficients for different drag functions are given for the following Mach numbers,

1.4; 1.5; 1.6; 1.8; 2; 2.2; 2.5,

we find that the average Mach number for the interval 1.39-2.5 is approximately 1.855, while the corresponding average speed is 632.23m/s.

In table 6.1, column C_{D8}, page 112 (ref. McCoy, R.L. "Modern Exterior Ballistics") it is shown that to the Mach number 1.9 corresponds a drag coefficient of approximately 0.3405.

Substituting the above values in formula (1.2), we obtain the corresponding average Siacci form factor,

$$i = 0.0014196 \frac{(647.58)^2}{647.58 - 240}(1.076) \cdot (0.3405) = 0.535 \ .$$

For the given projectile, considering that the mass of the bullet and the caliber are respectively 0.0097kg, and 0.0078m, we find that the ballistics coefficient (BC) is

$$c = 1000 \frac{i \cdot d^2}{m} = 1000 \frac{0.535 \cdot (0.00782)^2}{0.00973} = 3.363 \ .$$

Note that the form coefficient, 0.535, is somewhat different from the value 0.538 found in example 5, page 97-98 of the "Exterior Ballistics with Applications". The difference is due to the different approaches.

The adjusted ballistics coefficient that gives better results is 3.264.

Remark

Using PC program RangeC.Bas, it can be seen that both form coefficients give quite the same results (Results obtained using RangeC. bas are compared with the results displayed in table 2, page 369 of the book "Exterior Ballistics with Applications".

Siacci Ballistics Coefficients

In the same way are estimated the Siacci ballistics coefficients for the bullets presented hereafter. The ballistics coefficients are adjusted using the PC program Coeff.Bas.

- 5.56mm BRL-1
  (caliber 0.00569 m; mass, 0.004208kg; initial speed 975m/s);
  BC = 2.964.
- 5.56mm M193
  (caliber 0.00569, mass 0.00357kg, initial speed, 996.70m/s);
  BC = 5.360.
- 5.56mm M855
  (caliber 0.00569, mass 0.00402kg, initial speed, 948m/s);
  BC = 4.246.
- Bullet 5.45x39mm Russian
  (caliber 0.0056m, mass 0.00347kg, initial speed, 900m/s);
  BC = 3.801.
- Bullet 7.62mm M80
  (caliber 0.00782mass, 0.009534kg; initial speed, 856.5m/s);
  BC = 3.182.
- Sierra 0.277, 90 Gr. HP
  (caliber 0.0070, 0.00584kg; initial speed, 1024m/s);
  BC = 6.250.
- Sierra 0.277, 110 Gr. SP
  (caliber 0.0070, 0.00713kg; initial speed, 975.40m/s);
  BC = 4.108.
- Sierra 0.277, 130 Gr. SP
  (caliber 0.0070, 0.00843kg; initial speed, 930m/s);
  BC =3.322.
- Sierra 0.277, 190 Gr. HPBT
  (caliber 0.00782, 0.01232kg; initial speed, 759m/s);
  BC = 2.533
- 0.264 Norma BT Match, 139 Gr.
  (caliber 0.0067, 0.0090; initial speed, 936m/s);
  BC = 2.080.
- Barnes 80 Gr. RN
  (caliber 0.0070, 0.01167; initial speed, 777m/s);
  BC = 4.232.
- 7.62mm M80
  (caliber 0.00782, mass 0.0095kg, initial speed, 856.50m/s);
  BC = 3.182.
- 7.62mm M118 match
  (caliber 0.00782, mass 0.01129kg, initial speed,792.5m/s);

    BC = 2.674.
- 7.62mm M852 HPBT
  (caliber 0.00782, mass 0.0109kg, initial speed, 807.70m/s);
  BC = 2.945.
- 7.62mm M852 HPBT
  (caliber 0.00782, mass 0.01135kg, initial speed, 792.5m/s);
  BC = 2.677.
- 0.50 Ball M33
  (caliber0.012954, mass 0.0.0929kg, initial speed, 899m/s);
  BC = 1.907.
- 0.30 Ball M2
  (caliber 0.00782, mass 0.00973kg, initial speed 853.44m/s);
  BC = 3.264.

Remark

The above estimated values are approximate and need to be verified and corrected if the results obtained for the trajectory of flight are not acceptable. The adjustment can be done using the PC programs presented in this book, or by firing tests.

Example 1.2 Non-Siacci Form Coefficient, G1

For the 7.62mm Russian bullet mass 0.0079kg, find the form coefficient that corresponds to G_1 drag function. Speed interval, 735m/s-460m/s; corresponding Mach number interval, 2-1.4.
The Siacci form coefficient of 7.62mm bullet is known to be 0.56, BC = 4.116.
The solution is obtained using the table on page 112 (ref. McCoy, R.L. "Modern Exterior Ballistics, Schiffer Publishing, 1999".

Solution

Consider the Mach numbers in the interval 1.4-2, i.e. the sequence of Mach Numbers 1.4; 1.5; 1.6; 1.8; 2.

The average Mach number that corresponds to the projectile speed interval 735m/s-460m/s is 1.7, the average speed is 580m/s, while the corresponding G_1 drag coefficients is respectively 0.634.

Substituting the above values in formula (1.2) and solving for i_n we find respectively:

The G_1 form coefficient is $i_1 = 0.629$ (BC=0.3074).

1.2 Estimation of the Ballistics Coefficient of Bullets

The PC program COEFF.BAS (see the PC program at the end of this chapter), can be used to calculate the Ballistics Coefficient,

$$c = 1000 \cdot i \cdot d^2 / m,$$

employing the experimental data for the flight of projectiles in the Standard Atmosphere (Ref. G. Klimi, Exterior Ballistics with Applications", page 81, Xlibris, 2008), or the data from the Range Table of a given projectile.

COEFF.BAS (Ref. G. Klimi, Exterior Ballistics with Applications", page 414, Xlibris, 2008) can be easily modified for the experimental data in the non-standard atmosphere, or when the projectile characteristics are non-standard.

We assume that the origin of the coordinates of the Cartesian system is at the sea level.

Remarks

- The value of a ballistics coefficient "c" (hereafter denoted BC) might slightly change if we input in the PC program COEFF.BAS another guessed BC, or the value of the "error" but the change is insignificant.
- Sometimes the program might enter in a cycle. In this case, the values of the BC presented in the second column, alternates between two approximate values. For the value of BC we can chose one of those two displayed values. (See example 1.2.). We need to stop the program. To get a value of the BC we must again execute the program after changing the x-coordinate error.

- There might be cases where the program does not give an answer, and the message overflow does appear. In this case, interrupt the execution of the program. One reason might be that the input data do not correspond to a real shooting; for example, the departure angle does not correspond to the practical range of shooting.
- The BC estimated for any range depends on the error in x-coordinate we input in the program. If, for example, we want an accuracy of 1cm (0.01m) in the vertical direction we should input an error calculated using the formula:

$$error = \frac{0.01}{\tan\alpha_T}, \text{ or approximately, } error = \frac{0.01}{\tan\alpha_0}.$$

where α_0 is the angle of departure, and α_T is the terminal angle.

- The BC estimated using PC program Coeff.Bas need too be verified using the PC program RangeC.Bas, presented in chapter 3, or PC program AngleC.Bas, displayed in chapter 2;
- The BC can be estimated as well using the Program RangeC.Bas, by guessing the BC to let converge coordinates of the impact point to the measured coordinates of impact (and possibly to the terminal speed of the bullet), for a given departure angle and initial speed;
- The program Coeff.Bas can be modified for the projectile flight in a non-standard atmosphere.

Example 1.3

Use the Coeff.Bas to find the BC of a 122mm Russian projectile fired with initial speed of 885m/s if the departure angle is 20.75 degree and the range is 18,200 meters. Meteorological conditions are standard, the gun and the target are at the sea level.

Solution

Input a guessed initial value of BC, for example 0.2.

Input: The initial speed, 885m/s; the departure angle, 20.75; x-coordinate of target, 18200; largest range, 18200; error in x-coordinate, 1; number of steps, 10.

The computer will automatically find the value of the BC, 0.2519. The error in y-coordinate is 0.81.

The BC is saved in the file c:/koef.dat as well.

Note: The error in y-coordinate is around 0.81meter (81cm). If we want a smaller error then input a smaller "error in x-coordinate", i.e. instead of 1m we input 0.5m, and the "number of steps 1" (instead of 10). In this case, the BC remains the same.

Example 1.4

Find the BC for a projectile that is fired with initial speed 875m/s if the departure angle is 19 degree, and the range is 3496m.

Note. In Example 1 page 283 of "Exterior Ballistics with Applications" (Xlibris, 2008) we used a BC value of 3.6 that was not appropriate.

Solution

(a) To obtain a fast result when we execute the program Coeff.bas we round up the range to 3500m.

   Input: A guessed BC value of 3.5; the projectile speed 875; departure angle, 19; Corresponding range, 3500; largest range, 3500; error, 1; n = 100; enter.

   Output: The program will enter in a cycle and the value of the BC alternate between 3.6832 and 3.6833. The error oscillates between -12.31 and 6.125

   Interrupt the program: Click on the sign × on the right corner of the window coeff.bas.

(b) To find a more accurate value of the BC, we execute again the Coeff.bas inputting the guessed BC of 3.68, and all the other data, like in (a), but we must give to "the step" the value n =1. The result is 3.6798, and is valid for the range 3,500m. The y-error is 0.98m.

(c) We can avoid step (b), to find the BC that corresponds to the given range 3496m, inputting a guessed BC of 3.68; the projectile speed 875; departure angle 19, Corresponding range 3496; largest range 3496; error, 0.2; the step n =1;

The value of BC is 3.6856. The error in y-coordinate is around 0.12.

Exercise 1.5

Use Coeff.Bas to find the BC of the Russian 7.62mm bullet (Mod. 1943) that corresponds to the range 500 meters, if the bullet is fired with a speed 735m/s and the departure angle is 0.432 degree. Meteorological conditions are standard and the shooting is at the sea level. Find as well the form factor.

Answer: BC = 4.116; error 0.012m; form factor 0.56. (To find the form coefficient use the formula 1.1).

Remark

Seems that the form factor i = 0.56 usually given in the manuals of the 7.62mm Russian bullet, Mod. 1943, mass 0.0079kg, is obtained using the experimental data for the range 500 meter.

Employing the given form factor (i = 0.56) we can obtain approximate results that can be adjusted using experimental data for ranges 600-800 meters.

Exercise 1.6

Find the BC for the bullet 8x 57 SS Mauser K98k (Ref. Mori, E.) that corresponds to the range 500 meter, if the initial speed of the bullet is 755m/s and the departure angle is 0.286111 degree. Meteorological conditions are standard. **Answer**: 1.411.

Exercise 1.7

For the 174 grains, boat-tail Swiss bullet of caliber 0.30 the maximum range 4075.50m is obtained when the departure angle is 34.70 degree (Ref. Rinker, R. A. "Understanding Firearm Ballistics" p.266, New 6th Edition, Mulberry House Publishing). The initial speed of projectile is 792.48m. Find the Siacci Ballistics coefficient that corresponds to that max distance.
Answer BC = 3.2908.

Exercise 1.8

For the bullet caliber, 0.30 Ball M2 launched with speed 853.44m/s the maximum range 3,140 meter is obtained when the angle of departure is 32 degree (Ref. McCoy, R. L. "Modern Exterior ballistics", Figure 8.7, page 172, Schiffer Publishing Ltd., 1999). Bullet mass is 0.0097kg.
Find the corresponding Siacci Ballistics coefficient using Coeff.bas.
Answer. BC = 4.8439.

Remark

Using the same procedure and the data given by McCoy (see the above-referred graph), we find the following Ballistics coefficients for the bullet caliber 0.30 Ball M2 fired with a speed of 853.44m/s:

Angle = 32 degree, corresponding BC = 4.8439
Angle = 45 degree, corresponding BC = 4.8854
Angle = 60 degree, corresponding BC = 4.9775
Angle = 70 degree, corresponding BC = 4.9730
Angle = 85 degree, corresponding BC = 4.9790

Appropriate BC

For a given angle of departure, employing the above data, we can estimate an appropriate BC by interpolation.

Ballistics Coefficient as a Function of Departure Angle

Employing the data displayed above we find the ballistics coefficient of the given bullet as a function of the departure angle. For the departure angles greater than 30 degrees until around 86 degrees the ballistics coefficient is

$$c = 4.8694717 - 0.00701944 \cdot \alpha_0 + 0.00024886308 \cdot \alpha_0^2 - 0.000001781809 \cdot \alpha_0^3,$$

where α_0 is the departure angle.

Average Ballistics Coefficient

We can obtain as well an average ballistics coefficient, that can be used to estimate the elements of the trajectory when the projectile is launched with angles greater than around 30 degrees, but the results we will obtain using the average value will be not a good approximation.

Exercise 1.9

Find the BC of the 107mm French cannon if the projectile fired with initial speed 579m/s, at an angle of 7.5 degree will hit the ground at a horizontal range of 5220m. Find as well the form coefficient if the projectile mass is 16.380kg.
Answer. 0.4021; 0.575

Exercise 1.10

Find the BC of a 210mm mortar if the projectile fired with initial speed 213m/s, at an angle of 60 degrees will hit the ground at a horizontal range of 3460m. Find as well the form coefficient if the projectile mass is 78.70kg.
Answer. 0.3755; 0.670

Exercise 1.11

For the 7,62mm Russian bullet, mass 0.0079 kg, fired with initial speed 735m/s, use the PC program Coeff.Bas to find the ballistics coefficients that correspond to the following departure angles, corresponding ranges and speeds:

Departure Angle (Degree)	Range (meter)	Impact speed (m/s)
0.132	200	557
0.216	300	485
0.318	400	424
0.432	500	373
0.600	600	332
0.780	700	300
1.020	800	276
1.320	900	258
1.620	1000	244

Answer: 4.8945 (Input: error, 4), 4.5964 (error, 2); 4.4141, 4.132; 4.2960; 4.2206; 4.2965; 4.4235; 4.410.

Note 1. The answer depends on the error we enter. We should modify the input error, in order to have a desired output error.

Verify the results using RangeC.Bas. If the results are not acceptable, guess another coefficient using RangeC.Bas, or use again Coeff.Bas, but change the input error.

Note 2. The BC for range 100 meters cannot be obtained using Coeff. Bas. We can guess that ballistics coefficient using the RangeC.Bas, trying to match the calculated impact speed at 100 meters with the projectile speed 640m/s.

Using RangeC.Bas, we find that the best value of the BC for the horizontal range 100 meters is 4.4.

1.3 Determination of an Appropriate Ballistics Coefficient

The problem of determining the ballistics coefficient of a bullet is essential in the practice of shooting. Since the BC is a function of the projectile speed, and the speed of the projectile changes during the flight, we need to find the functional variation of the BC with the projectile speed.

Using the software technology (for example using the PC program Coeff.Bas, or the Siacci method (Ref. "Exterior Ballistics with Applications", Example 3, p.242) it is easy to determine the BC as a function of bullet speed, but the equipment (wind tunnels, radars) we need to use to measure the projectile speed and the drag coefficients are very expensive and usually available in ballistics research laboratories of the armies and some manufacturing companies.

Moreover a variable BC is not easy to be used in the practice of shooting with small arms to obtain the initial firing data, unless we have time to find the initial data of shooting (using a PC), to set up the sight or change the aiming point and then fire.

For practical purposes, the Exterior Ballistics uses a fixed BC valid for an interval of bullet speed, or for a given interval of horizontal range of shooting.

The BC of some bullets is determined practically for short distances, usually 100 meters. The use of such measured ballistics coefficients, in general, does not give satisfactory results to construct the Range Tables, or to study the projectile trajectory.

The problem we come across is **"Which is an appropriate BC for a given bullet and how to determine it?"**

We will try to give an answer to that question, considering as illustration the caliber 7.62mm, M59 Ball bullet, fired with an initial speed of 853.44m/s (2800 feet/s), Ref. Rinker, R.A., "Understanding Ballistics of Arms", p. 269, 6th. Ed., Mulberry House Publishing, 2005.

We assume that the muzzle of the gun and the center of the target are at the same horizontal plane, at the sea level.

We consider the data presented in the above reference as experimental data. Based on those data and using the PC program Coeff.Bas we have estimated the ballistics coefficient and the terminal angles for ranges 200-1500m. Those data are presented below.

Range	Departure angle	Terminal Angle	BC
200	0.0900	-0.09700	3.2000
300	0.1463	-0.18256	3.600
400	0.2083	-0.27506	3.4500

500	0.2817	-0.40263	3.4085
700	0.4666	-0.77634	3.3400
800	0.5917	-1.06292	3.3700
1000	0.9117	-1.84184	3.3900
1500	2.2331	-4.85953	3.4776

Now, let us estimate the vertical deviation of the given bullet from the center of the target if we consider the BC equal to a particular value, let say 3.60, i.e. equal to the BC value that corresponds to the target horizontal range 300m.

The error in the horizontal range (Ref. formula (6.1.4)), "Exterior Ballistics with Applications", p. 304), is

$$\Delta x = -(1 - \frac{\tan \alpha_0}{\tan |\alpha_T|}) \frac{dc}{c} x_T , \qquad (1.3.1)$$

while the error in the y-coordinate is

$$\Delta y = \tan |\alpha_T| \Delta x = -(\tan |\alpha_T| - \tan \alpha_0) \frac{dc}{c} x_T . \qquad (1.3.2)$$

Estimation of Vertical Errors

I. Considering BC = 3.60, we find the following vertical deviations:

- For the horizontal range 500 meters, substituting in (1.3.2) we obtain the vertical deviation from the center of the target located 500 meters from the muzzle:

$$\Delta y = -[\tan(0.40326) - \tan(0.2817)] \frac{(3.60 - 3.4085)}{3.4085} (500) = -0.06m .$$

In the same way, we find:

- For the horizontal range 800 meters, the vertical deviation is $\Delta y = -0.45m$.

- For the horizontal range 1000 meters the vertical deviation is $\Delta y = -1.01m$.

II. Considering BC = 3.4085, i.e. the BC value that corresponds to the horizontal range 500 meters we find the following vertical deviations from the center of the target:

- For the horizontal range 300 meters, $\Delta y = 0.010m$;
- For the horizontal range 400 meters, $\Delta y = 0.006m$;
- For the horizontal range 700 meters, $\Delta y = -0.078m$;
- For the horizontal range 800 meters, $\Delta y = -0.075m$;
- For the horizontal range 1000 meters, $\Delta y = -0.089m$,
- For the horizontal range 1500 meters, $\Delta y = -1.372m$.

Verification

Using the PC program RangeC.Bas, we can see that the vertical deviations are correctly estimated.

Indeed, for the horizontal range 800 meters **RangeC.Bas** give the following results:
Input: Departure angle, 0.5906; Departure Speed, 853.44; BC = 3.4085.
Output: Range 796m, Impact Angle, -1.066703.
For the vertical deviation, we have:

$$\Delta y = (800 - 796) \cdot \tan(-1.066703) = -0.0745$$

Comments

- As we can see from the above example, if we use the BC measured for a distance 500 meters, the vertical deviations from the center of the target are much smaller than the vertical deviations the bullet will have if we consider a ballistics coefficient measured for a smaller distance (let say for a distance 100 meters).
- The BC determined for the range 500m gives acceptable results for the effective range of small arms, until 600-800 meters.

- It can be proved that the vertical deviations will be much greater if we consider a BC measured for a distance around 100 meters, or less.

Some ballisticians measure the BC for distances around 100 meters. That (short range) calculated BC gives good results for small ranges, but for relatively large distances, the errors in the vertical direction are significant.

It is recommended to measure the BC for horizontal ranges 500m, 800m, or sometimes 1000m, firing on vertical sheets.

- For larger distances, the measurement of the BC should be done for ranges 1500, 2000, 3000, or so on dependence of the maximum range of fire. For large distances (over 1,500), in field conditions, we have difficulty to find the points where the bullets hit the ground, mostly because of ricochets. One solution might be to fire on a frozen river (lake), in order to have the possibility to locate the center of the impact of the bullets on the ground.
- The BC found for the horizontal range can be used for the inclined ranges until around 1000 meters.
- Since in practice it is not always possible to have an ideal standard atmosphere, or to have practically such weather conditions that the atmosphere could be considered standard, the experiments should be performed in a non-standard atmosphere, but in absence of wind.

1.4 Firing Tests to Determine the Ballistics Coefficient

Determination of the BC through practical measurements is the best way to estimate an appropriate BC for projectiles fired by small arms.

In absence of wind tunnels, or radars, the exterior ballistics uses firing tests that can be practically performed by any skilled shooter using a chronograph to measure the time of flight of the bullet to the target, and the bullet speed.

Based on the results obtained analytically in the above section, we will consider a BC that is measured for relatively large horizontal ranges, usually 500 meters.

Experimental Determination of the Ballistics Coefficient

We consider that the initial speed, departure angle, the horizontal range, and the time of flight are measured experimentally with a satisfying accuracy. The atmosphere is standard or close to it, and the shooting is performed at the sea level. To determine the BC we can use:

- The Siacci's Method.
- The approximate formula (3.2.12), ref. "Exterior Ballistics with Applications, p.128.
- The approximate formula (3.3.4), ref. "Exterior Ballistics with Applications, p.137.
- We can use the formula (3.5.2), page 162, Ref. "Exterior Ballistics with Applications" when we know the drop of the projectile fired with departure angle 0 degree.
- The PC program **Coeff.bas** (Universal method, valid for any firearm).
- The PC program **RangeC.Bas,** or **AngleC.Bas**.

The BC estimated using one of the above methods could be employed to study the projectile trajectory using any of the methods: Siacci's method, approximate method that involves the approximate formulas, or the PC programs.

It is obvious that to have better results in estimating the elements of the trajectory of flight the BC obtained using one of the above methods should be used with the corresponding method.

Siacci Method

We assume that we are able to measure, in practice of shooting with a given small arm and a given bullet, the time of flight to the target, the angle of departure and of course the horizontal range to the target. We consider a standard atmosphere. The muzzle and the target are at the same vertical distance from the ground. Their coordinates can be considered equal to zero.

Consider the Siacci formulae we have introduced in section 5.2 of "Exterior Ballistics with Applications":

$$x = -\frac{1}{Bg}[D(u) - D(v_0)] \qquad (1.4.1)$$

and

$$t = -\frac{1}{Bg\cos\alpha_0}[T(u) - T(v_0)], \qquad (1.4.2)$$

where

$$B = \frac{ch(\overline{y})}{3g\sqrt{\cos\alpha_0}}, \quad h(\overline{y}) = (\frac{289.08 - 0.006328\overline{y}}{289.06})^{4.4}, \quad \overline{y} = 2y_{max}/3 \quad (1.4.3)$$

$$D(u) = u + 240 \cdot \ln(u - 240), \quad D(v_0) = v_0 + 240 \cdot \ln(v_0 - 240), \quad (1.4.4)$$

and

$$T(u) = 240 \cdot \ln(u - 240), \quad T(v_0) = 240 \cdot \ln(v_0 - 240). \qquad (1.4.5)$$

Since for small arms the departure angles are very narrow in the first formula of (1.4.3) we can consider the cosine of the departure angle equal to one). Thus, that formula can be written in the form:

$$B = \frac{ch(\overline{y})}{3g}.$$

We find B as a solution of the above system of equations (1.4.1) and (1.4.2). Indeed, eliminating B from equations (1.4.1) and (1.4.2) we obtain

$$\frac{T(u) - T(v_0)}{t \cdot \cos(\alpha_0)}x = D(u) - D(v_0) \qquad (1.4.6)$$

Solving for "u" employing a graphing calculator TI-83+, and then substituting the obtained value of "u" in one of the equations (1.4.1) or (1.4.2) we find B, and then considering (1.4.3), it is easy to find the ballistics coefficient of the given projectile.

Note that with this method it is not necessary to determine experimentally the terminal speed of the projectile, though a measured terminal speed can simplify the calculations and can serve to verify the accuracy of the obtained BC.

Example 1.12

A Russian bullet caliber 7.62mm mass 0.0079kg, fired with a measured initial speed of 735m/s at an angle 0.432 degree, hits the center of the target at the horizontal range 500 meters from the muzzle. The time of flight is 0.97 seconds. The tests are performed at the sea level in standard conditions. Find the ballistics coefficient.

Solution

Substituting in (1.4.6), the given data we have:

$$\frac{T(u)-T(735)}{0.97 \cdot \cos(0.432)}(500)=D(u)-D(735)$$

Substituting (1.4.4) and (1.4.5) in the above equation and solving the obtained equation, using a graphing calculator TI-83+ we find:

$$u = 373.03 m/s.$$

Substituting the above value in (1.4.1), and considering the first equation of (1.4.4) we find that BC is

$$c = 4.064.$$

Approximate Formula

We will solve the problem displayed in example 1.12 using the formula (3.2.12), ref. "Exterior Ballistics with Applications", p.128:

$$x = 240t + \frac{(v_{x0} - 240)}{b}(1 - e^{-b \cdot t}), \qquad (1.4.7)$$

where

$$b = ch(\bar{y})/3. \qquad (1.4.8)$$

Substituting in (1.4.7), we obtain the following equation:

$$500 = 240(0.97) + \frac{735\cos(0.432) - 240}{b}(1 - e^{-0.97b}).$$

The solution of the above equation using a graphing calculator TI-83+ is $b = 1.35462$.

Substituting the value of "b" in (1.4.8) and solving the obtained equation, we find that the ballistics coefficient is $c = 4.064$.

Remark

As we see, the ballistics coefficients obtained using the Siacci method or the approximate formula, are the same.

The value of ballistics coefficient $c = 4.116$ of the given bullet obtained using the form factor 0.56 is slightly different from the value obtained above, but gives a slightly different terminal speed.

Nevertheless, considering the analyses performed in section 1.3, it follows that the elements of projectile trajectory obtain using $c = 4.064$ (form factor 0.553) instead of $c = 4.116$ (form factor 0.56), are practically the same.

Example 1.13

The caliber 7.62mm, M-59 Ball bullet, fired with an initial speed of 853.44m/s (2800 feet/s) at an angle 0.04 degree, hits the target located at the horizontal range 100 meters from the muzzle. Find the BC of the bullet that corresponds to the range 100 meters at the sea level. Consider a standard atmosphere (Ref. Rinker, R.A., "Understanding Ballistics of Arms", p. 269, 6th. Ed., Mulberry House Publishing, 2005).

Solution

Since we do not know the time of flight to the target, we can use the equation (3.3.4) page 137 of "Exterior Ballistics with Applications":

$$y = x \cdot \tan \alpha_0 - \frac{g}{2 \cdot v_0^2} \cdot x^2 - \frac{gb(v_0 - 240)}{3v_0^4} x^3 - gb^2(v_0 - 240) \frac{v_0 - 320}{4v_0^6} x^4 \,. \quad (1)$$

Substituting $y = 0$, we have

$$\tan \alpha_0 - \frac{g}{2 \cdot v_0^2} \cdot x_T - \frac{gb(v_0 - 240)}{3v_0^4} x_T^2 - gb^2(v_0 - 240) \frac{v_0 - 320}{4v_0^6} x_T^3 = 0 \,, \quad (2)$$

where

$$b = ch(\bar{y})/3 \,, \qquad (4)$$

and x_T is the abscissa of the target.
    Substituting in equation (2),

$$\alpha_0 = 0.04 \,, \quad v_0 = 853.44 \,, \quad x_T = 100 \,,$$

we obtain the following second-degree equation:

$$2.493 \times 10^{-5} - 3.7799 \times 10^{-5} b - 2.07625 \times 10^{-6} b^2 = 0 \,.$$

The solution of the above equation is

$$b = 0.63726 \,.$$

Substituting in (4) and considering that $h(y) = 0$, we find that the ballistics coefficient is

$$c = 3 \cdot b = 3(0.4328) = 1.912.$$

Remark

The BC obtained using the above approach, is approximate. It needs to be adjusted using the PC programs.

Exercise 1.14

The caliber 7.62mm, M-59 Ball bullet, fired with an initial speed of 853.44m/s (2800 feet/s) at an angle 0.2812 degree, hits the target at the horizontal range 500 meters from the muzzle. Find the BC of the bullet that corresponds to the range 500 meters at the sea level.

Consider a standard atmosphere, and the shooting is at the sea level. (Ref. Rinker, R.A., "Understanding Ballistics of Arms", p. 269, 6th. Ed., Mulberry House Publishing, 2005).

Answer: BC = 3.665.

Example 1.15

At the point with abscissa 320 meters the drop of the caliber 0.257, Sierra 117 grain bullet, fired with an initial speed of 883.92m/s at an angle zero degree, is 0.7965m. (Ref. http://exteriorballistics. com/ebexplained/5th/50.cfm).

Find the BC of the bullet that corresponds to $x = 320$ meters if the shooting is at the sea level. Consider a standard atmosphere. Use the PC program Coeff.Bas.

Solution

Input: Initial BC = 4; projectile speed, 883.92; departure angle, 0; x-coordinate of target, 320; y-coordinate of target, 0; y-coordinate of gun, 0.7965, error in x-coordinate is 10; number of steps, 10.
Output: BC = 3.3170.

EXTERIOR BALLISTICS OF SMALL ARMS

The estimated BC can be adjusted using PC program RangeC.Bas. The corrected BC is equal to **3.145**.

Note. Since the origin of the Cartesian system of coordinates is at the sea level, we consider that the muzzle of the rifle is 0.7965 meters over the sea level. The projectile line of fire is horizontal. The bullet point of impact is on the ground (y =0) at the point with abscissa 320.

When the firing tests are done in a non-standard atmosphere, we convert the experimental data into standard data, using the method shown in section 6.2 of the book "Exterior Ballistics with Applications".

Exercise 1.16

At the point with abscissa 503 meters (550 yards), the drop of the caliber 0.257, Sierra 117 grain bullet, fired with an initial speed of 883.92m/s at an angle zero degree, is 2.2677m. (Ref. http://exteriorballistics. com/ebexplained/5th/50.cfm).

Find the BC of the bullet that corresponds to the horizontal range x = 503 meters at the sea level. Consider a standard atmosphere. Use the PC program Coeff.Bas, and PC program RangeC.Bas.

Answer: BC= 3.3026

Note. The BC = 3.3026 can be used to obtain the elements of the trajectory of flight for horizontal ranges until 500-800 meters (550-850) yards.

1.5 Projectile Drop, Estimation of Departure Angle

The Exterior Ballistics of small arms has some characteristics that allow a marksman to simplify the practice of firing with sight small arms.

Hereafter (sections 1.5-1.7) there are formulated some characteristics of trajectories of the small arms (Ref. Okunev, B. H, "Fundamentals of Ballistics", page 239-241, Vol.1, Book 2, Moscow, 1943).

Those characteristics are obtained for the flight of projectiles in presence of air resistance using the Siacci approach.

General Characteristics of a Bullet Trajectory

The drop of a projectile, at a given time and location during the flight, is the perpendicular distance of the bullet location measured from the line of departure (Figure 3, Figure 4). In curvilinear coordinates, the drop is denoted by $\bar{y}$ (see, G. Klimi, "Exterior Ballistics with Applications", p. 148, Xlibris, 2008).

For the inclined fire (uphill, or downhill shooting, or anti-aircraft fire) the effective range of the sight arms is relatively short, and it is on the ascending part of the trajectory. The departure angles are large, but the curvature of the trajectory is relatively small, and deviates not significantly from the departure line, and so that the use of the Siacci approach that involves the pseudo speed is suitable for the study of the inclined fire.

For such "inclined" trajectories, we can apply the following properties:

- For any two projectiles that have the same ballistics coefficients $c_2 = c_1 = c$ (same form coefficient i) and are fired with the same initial speed v_0, but with different angles of departure (respectively α_{01} and α_{02}) the drops of the projectiles at the points on the trajectories with the same abscissa $x_2 = x_1 = x$ (Figure 3) are related by the equation:

$$\frac{\bar{y}_2}{\bar{y}_1} = \frac{\cos^2(\alpha_{01})}{\cos^2(\alpha_{02})}, \qquad (1.5.1)$$

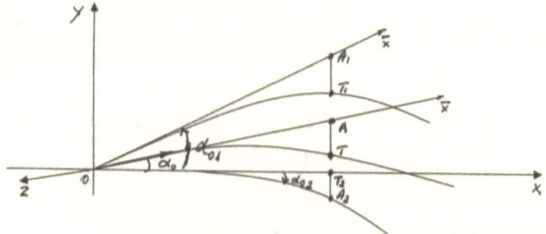

Figure 3—Siacci's Trajectories

TOA_1, TOA_2, and TOA—Departure Angle;
A_1T_1, A_2T_2, and AT—Projectile Drop.

The property holds also if at least one of the trajectories that is near to the horizontal line has the absolute value of the departure angle less than 15 degree, i.e. $|\alpha_0| \le 15°$.

Comment

Note that the property (1.5.1) is valid assuming that the ballistics coefficient remains constant when the departure angle increases or decreases.

In fact, we can consider the ballistics coefficient constant when the changes in the departure angle from zero are relatively small. For relatively big changes, the ballistics coefficient does not have the same value as it has when the departure angle is zero.

Hence, assuming a ballistics coefficient constant and independend from the departure angle, we obtain approximate values that are acceptable within the effective range of small arms.

For that reason, even the PC programs that use a fixed ballistics coefficient (Rangec.Bas, AngleC. Bas, etc.) give approximate values for the elements of the trajectories when the departure angle is large, as it is the case of the inclined fire.

For the same reason, even the following Rifleman's Rules are approximate.

On the other hand, for narrow departure angles, the ballistics coefficient remains equal to the ballistics coefficient estimated for departure angle zero, within the relatively small ranges of the small arms shooting. Thus, the estimation of the departure angle based in the following procedure is accurate.

Estimation of the Departure Angle

For two different trajectories of the same projectile, with departing angles close to zero (cosine approximately one), formula (1.5.1) shows that the drops that corresponds to the same x-coordinate are approximately the same,

$$\overline{y}_2 = \overline{y}_1. \tag{1.5.2}$$

Using (1.5.2), **we can find the angle of departure α_0 needed to zero the projectile at any relatively short horizontal range x_T, if we know the drop, $\overline{y}_T$, of the same projectile, but fired with a departure angle equal to zero, at the point with abscissa x_T.**

Indeed, the departure angle needed to zero the bullet at the horizontal range x_T satisfies the equation:

$$\tan(\alpha_0) = \frac{\overline{y}_T}{x_T},$$

where $\overline{y}_T$ is the drop of projectile (fired horizontally) at the point with abscissa $\overline{x}_T$.

Solving the above equation, we find that the departure angle is

$$\alpha_0 = \tan^{-1}(\frac{\overline{y}_T}{x_T}). \tag{1.5.3}$$

To find the departure angle (in radian) of small arm projectiles, within the effective range, we can use the approximate formula:

$$\alpha_0 = \frac{\overline{y}_T}{x_T},$$

where the drop and the x-coordinate of target are expressed in the same units.

The departure angle in MOA (Minute of Angle) is

$$\alpha_0 = \frac{\overline{y}_T}{x_T} \cdot \frac{10,800}{\pi}$$

The angle of sight

If the sight height of the gun is "h_s" then the angle that the line of sight (LOS) forms with the horizontal line (Figure 1) is

$$\alpha_h = \tan^{-1}(\frac{-h_S}{x_T}) \tag{1.5.4}$$

The angle of sight is given by the formulas:

$$\alpha_S = \alpha_0 - \alpha_h,\tag{1.5.5}$$

or

$$\alpha_S = \tan^{-1}(\frac{\overline{y}_T}{x_T}) + \tan^{-1}(\frac{h_S}{x_T}).\tag{1.5.6}$$

The following examples are solved using the property (1.5.1), and (1.53).
For relatively long horizontal ranges, to find the drop of the projectile at the "zeroing" point we use (1.5.1).
Formula (1.5.6) can be written in the following approximate form:

$$\alpha_S = (\frac{\overline{y}_T}{x_T} + \frac{h_S}{x_T}) \cdot \frac{10,800}{\pi},\tag{1.5.7}$$

where the angle of sight is expressed in MOA.

Example 1.17

The drop of a Sierra NATO bullet caliber 0.30, 168 grain HPBT, fired horizontally, at the sea level, with initial speed 807.72m/s at the horizontal range 548.60 meters (600 yards) is 3.213 meters.

(a) Find the angle of departure needed to zero the gun at a horizontal range of 548.60 meters (600 yards) from the gun.
(b) Find as well the angle of sight that the line of bullet departure forms with the line of sight if the sight height is 0.0381m (1.5 inches).

Solution

(a) From (1.5.2), since the x-coordinate (548.60m) is relatively small, we can consider that the drop of the projectile, fired with an angle α_0 that zeros the bullet at the horizontal range 548.60m is equal to 3.213 meters.

The angle of departure α_0 that zeros the projectile at the range 548.60m, is

$$\alpha_0 = \tan^{-1}(\frac{\overline{y}_T}{x_T}) = \tan^{-1}(\frac{3.213}{548.60}) = 0.335572° = 20.134 MOA \cdot$$

We obtain the same result if we use the approximate formula

$$\alpha_0 = \frac{\overline{y}_T}{x_T} \cdot \frac{10,800}{\pi} = \frac{3.213}{548.60} \cdot \frac{10,800}{\pi} = 20.134 MOA \cdot$$

(b) For the angle of sight that the line of sight (LOS) forms with the horizontal line at the horizontal range 548.60m we can write:

$$\tan(\alpha_h) = \frac{s}{x} = \frac{-0.0381}{548.60} = -0.00006945.$$

Hence,

$$\alpha_h = \tan^{-1}(-0.00006945) = -0.003979° .$$

The angle of sight is

$$\alpha_S = \alpha_0 - \alpha_h = 0.335572 - (-0.003979) = 0.33955° = 20.37'.$$

Thus,

$$\alpha_S = 20.37 \ MOA .$$

Note. 1 MOA (Minute of Angle) is equal to an angle of 1 minute.

Example 1.18

A Russian bullet, caliber 7.62mm, mass 7.9 gr. is fired with an initial speed of 735m/s from a rifle zeroed at the range 500 meters. The angle of departure is 0.432 degree.

The average height of the trajectory at the point with abscissa 300 meters is 1.20 meters.

(a) Find the angle of departure needed to zero the rifle at the horizontal range 300 meters.
(b) Find as well the angle of sight considering a sight height of 5 centimeters.

Solution

(a) The drop of projectile at the point with abscissa 300m is:

$$\bar{y}_T = 300\tan(0.432) - 1.14 = 1.122m .$$

The drop of the gun zeroed at 300 meters is equal to the above estimated drop.
The angle of departure needed to zero the rifle at the horizontal range 300m is

$$\alpha_0 = \tan^{-1}(\frac{1.122}{300}) = 0.214° = 12.86 \ MOA .$$

Note. The above result $\alpha_0 = 0.214°$ obtained using (1.5.1), is approximately equal to the value $\alpha_0 = 0.216°$ given in the range tables of the 7.62 mm Russian rifle.

The difference is 0.10 MOA.

(b) The angle of sight needed to zero the rifle at 300 meters is

$$\alpha_S = \tan^{-1}(\frac{\bar{y}_T}{x_T}) + \tan(\frac{h_S}{x_T}) = \tan^{-1}(\frac{1.122}{300}) + \tan^{-1}(\frac{0.05}{300}) = 0.224° .$$

or

$$\alpha_S = 13.43 \ MOA .$$

1.6 Inclined Fire and the Rifleman's Rule

The sight arms are used in uphill or down hill shooting, against the parachutists or helicopters, as well as in hunting on mountains. In such firing situations, to make shooting practical and accurate we will study some characteristics of sight arms, that allow us to use the range tables that are constructed for horizontal shooting, in order to set up the sight of a firearm for the inclined range.

Hereafter is shown a method we can use to estimate the super elevation angle, and based on that to set up the sight of the firearm to hit the target located at a given inclined range. The method is based on the geometric relations between elements of two trajectories (Figure 4) in presence of air drag. The formulas can be easily obtained using the theorem of sine. The results are valid as well for some cannons, or other infantry arms.

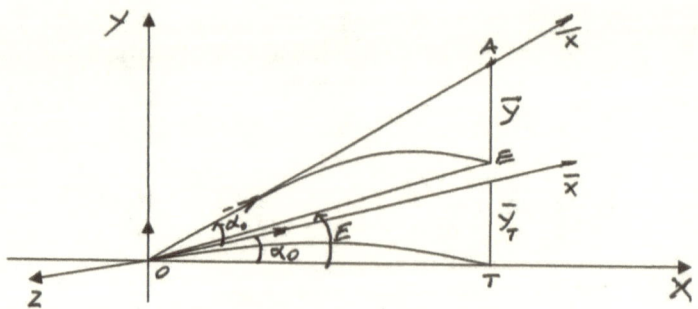

Figure 4—Projectile Drop

OA, and OB—Line of Departure; TOB, and TOA—Departure angle
OT—Horizontal Rangge, OE—Inclined Range;
TOE—Angle of Site; EOA—Super Elevation Angle;
AE = $\bar{y}$, projectile drop; BT = $\bar{y}_T$, projectile drop.

Consider a trajectory of a projectile fired with initial speed v_0, departure angle α_0 (in degree), where $|\alpha_0| \leq 15°$. The impact point is at the horizontal range x_T (Figure 4). Consider as well another trajectory of a projectile that have the same ballistics coefficient and the same initial speed as the first trajectory, but it is fired with departure angle $(\bar{\alpha}_0 + E)$, where $\bar{\alpha}_0$ is the super elevation angle, and E is the angle of site. We consider that the point of impact is on the inclined plane at the slant range,

$$D = \frac{x_T}{\cos(E)}, \tag{1.6.1}$$

that has the same abscissa, $x_i = x_T$. The projectile is zeroed at the inclined range determined by (1.6.1).

The second trajectory might be the trajectory of the same projectile, or of another projectile, that has the same BC and the same initial speed as the trajectory of the first projectile.

The Adjusted Rifleman's Rule

For those two trajectories it can easily be proved (Ref. Okunev, B. H, "Fundamentals of Ballistics", page 241, Vol.1, Book 2, Moscow, 1943) that:

- The inclined angle of fire $\bar{\alpha}_0$ that zeroes the gun at the given inclined range D is given by the equation:

$$\sin(2\bar{\alpha}_0 + E) = \sin(2\alpha_0)\cos(E) + \sin(E). \tag{1.6.2}$$

Using (1.5.1), for the drop $\bar{y}_E$ of the projectile fired on the inclined plane, at the given inclined range D, we have:

$$\bar{y}_E = \frac{\cos^2(\alpha_0)}{\cos^2(\bar{\alpha}_0 + E)}\bar{y}_T, \tag{1.6.3}$$

where

$$\bar{y}_E = x_E[\tan(\bar{\alpha}_0 + E) - \tan(E), \quad \bar{y}_T = x_T\tan\alpha_0, \tag{1.6.4}$$

and $\bar{y}_T$ is the drop of the projectile fired on the horizontal plane.

We will call the set of equation (1.6.2)-(1.6.4) the Adjusted Rifleman's Rule.

Rifleman's Rule

Employing the formula for the sine of sum, we can write (1.6.2) in the following form:

$$\sin(2\bar{\alpha}_0) \cdot \cos(E) + \cos(2\bar{\alpha}_0) \cdot \sin(E) = \sin(2\alpha_0) \cos(E) + \sin(E) \quad (1.6.5)$$

For relatively narrow angles, we can assume that:

$$\cos(2\bar{\alpha}_0) \approx 1, \text{ and } \cos(2\alpha_0) \approx 1.$$

Substituting the above approximate relations in equation (1.6.5), we find that the super elevation angle, $\bar{\alpha}_0$ that zeroes the firearm at the inclined range D, is approximately equal to the departure angle α_0 that zeroes the firearm to the horizontal range, determined by the equation:

$$x_T = D \cos E . \quad (1.6.6)$$

Thus, we can write:

$$\bar{\alpha}_0 \approx \alpha_0 . \quad (1.6.7)$$

For narrow angles, from (1.6.3), we find that the drop of the projectile that corresponds to the inclined impact point is approximately

$$\bar{y}_E = \frac{\bar{y}_T}{\cos^2(E)}, \quad (1.6.8)$$

where $\bar{y}_T$ is the drop of the projectile fired with departure angle α_0.

The result expressed by equation (1.6.7) is known as the **Rifleman's Rule**.

The **Rifleman's Rule** is obtained using the ideal parabolic trajectory of the projectile flight in absence of drag, (Ref. R.L. McCoy, "Modern Exterior Ballistics", p.48, Schiffer Publishing 1999; Ref. W. T. McDonald, "Inclined Fire", June 2003, www.exteriorballistics.com).

The **"Rifleman's Rule" as it is actually used in everyday practice of firing with sight arms need to be corrected.**

A more accurate Rifleman's rule is represented by the set of equations (1.6.6), (1.6.7), and (1.6.8) that are based on the formula (1.5.1). Those equations give a better estimation for the drop of the projectile that corresponds to the inclined fire.

The drop of a projectile that corresponds to the inclined fire can be estimated using (1.6.8), i.e.

$$\overline{y}_E = \frac{\overline{y}_T}{\cos^2(E)}, \tag{1.6.9}$$

To hit the target located at the inclined range D, the rifleman should find the projection of the target on the x-axis, i.e.

$$x_T = D\cos(E) \tag{1.6.10}$$

The angle of sight (in degree) that zeroes the firearm at the inclined range D and the angle of sight (in degree) that zeroes the firearm at the horizontal range x_T are respectively (See Figure 1, and Figure 2):

$$\alpha_{SD} = \overline{\alpha}_0 + \frac{180}{\pi}\frac{h_S}{D} = \alpha_0 + \frac{180}{\pi}\frac{h_S}{x_T}\cos(E), \tag{1.6.11}$$

and

$$\alpha_{S0} = \alpha_0 + \frac{180}{\pi}\frac{h_S}{x_T}, \tag{1.6.12}$$

where h_S is the sight height. In (1.6.1) we have considered the fact that $\overline{\alpha}_0 \approx \alpha_0$.

Note that in the formulae (1.6.11) and (1.6.12) we have used the fact that for small angles we can approximate the tangent of that angle with the angle itself expressed in radians. The ration $180/\pi$ transforms the radians into degrees.

Using (1.6.11) and (1.6.12), we find that the respective angles in MOA are:

$$\alpha_{SD} = 60 \cdot \bar{\alpha}_0 + \frac{10800}{\pi} \frac{h_s}{\pi} \cos(E), \qquad (1.6.13)$$

and

$$\alpha_{s0} = 60 \cdot \alpha_0 + \frac{10800}{\pi} \frac{h_s}{x_T}. \qquad (1.6.14)$$

Since $\bar{\alpha}_0 \approx \alpha_0$, the difference (in MOA) between the angle of sight for the inclined range and the angle of sight for the horizontal range is

$$\alpha_{SD} - \alpha_{S0} = \frac{10800}{\pi} \frac{h_s}{x_T} (\cos(E) - 1). \qquad (1.6.15)$$

Because the cosine of an angle is always less than 1, the difference (1.6.15) is negative. The inclined angle of sight (in MOA) that zeroes the firearm at the inclined range D is

$$\alpha_{SD} = 60 \cdot \alpha_{S0} + \frac{10800}{\pi} \frac{h_s}{x_T} (\cos(E) - 1) \qquad (1.6.16)$$

The "correction quantity" (1.6.15) is insignificant for angles of sights until around 30 degree and horizontal ranges 100 meters or over.

Thus, to set up the sight for the inclined range D, we just need to set up the sight for the horizontal range using (1.6.14), and then we can use that angle of sight to fire, pretending that the inclined range of shooting D is equal to the horizontal range x_T.

Comment

The super elevation angle and the projectile drop at the inclined range are estimated more accurately using equations (1.6.2) and (1.6.3).

We note also that the adjusted rifleman's rule, though more accurate than the rifleman's rule, is still an approximate rule for large departure angles or large angles of sight, since the condition of equal ballistics coefficient of two trajectories is not satisfied. We should avoid their use for relatively large distances of shooting.

Anyway, we will use the Rifleman's rules for short firing distances, especially for shooting with small firearms, assuming that the ballistics coefficient does not change with the angle of sight.

For illustration, see the following example 1, 19, example 2.25 section 2.5, and the example 2.26, section 2.6.

Example 1.19 Employing the Adjusted Rifleman's Rule

The Sierra caliber 0.257", 117 grain, SPT is fired horizontally at 883.92m/s (2900 feet per second. The drop of the bullet at the point with abscissa 503m (550 feet) is 2.284m (89.92 inches).

(a) Use the formula (1.5.3) to find the departure angle and the sight angle needed to hit a target located at the inclined range (581m) at the point with x-coordinate equal to 503m if the angle of site is 30^0.
(b) Find the drop of the projectile at the given inclined range.
(c) The projectile ballistics coefficient is found to be 3.360. Verify the results obtained in (a) and (b) using the PC program RangeC.Bas.
(d) Estimate the correction term given by (1.6.15) if the sight height is 0.0381m (1.5 inches).

We assume that the ballistics coefficient of the bullet does not change with the departure angle.

Solution

a. The departure angle to zero the firearm at the horizontal range 503 meters is

$$\alpha_0 = \tan^{-1}(\frac{\overline{y}_T}{x_T}) = \tan^{-1}(2.284 / 503) = 0.260164° = 15.61 \; MOA.$$

Employing (1.6.2) for the departure angle needed to hit the target at the inclined range 581 meters we can write:

$$\sin(2\bar{\alpha}_0 + 30) = \sin(2 \cdot 0.260164)\cos(30) + \sin(30) = 0.507865.$$

Hence, we find that the super elevation angle of fire is

$$\bar{\alpha}_0 = 0.260850° = 15.651\ MOA, \tag{1}$$

The super elevation angle is practically the same as the departure angle that corresponds to the horizontal range 503m. In fact, the super elevation angle is somewhat greater than the departure angle.

The angle of sight (in MOA) that zeroes the firearm at the inclined range 581 meters is

$$\alpha_{SD} = 60\bar{\alpha}_0 + \frac{10800}{\pi}\frac{h_S}{x_T}\cos(E) = 15.651 + \frac{(0.0381) \cdot 10800}{(503)\pi}\cos(30) = 15.88MOA \cdot$$

The angle of sight (in MOA) that zeroes the rifle at the horizontal range 503 meters is

$$\alpha_{SO} = \alpha_0 + \frac{h_S}{x_T}\frac{10800}{\pi} = 15.61 + \frac{(0.0381) \cdot (10800)}{\pi(503)} = 15.87MOA.$$

Comparing the above results, we see that we have to set up the firearm sight considering as if we are firing at the inclined range of 503 meters instead of the real inclined range 581 meters.

b. Employing (1.5.1), we find that the drop of the projectile at the given inclined range is

$$\bar{y}_E = \frac{\cos^2(\alpha_0)}{\cos^2(\bar{\alpha}_0 + E)}\bar{y}_T = \frac{\cos^2(0.260164)}{\cos^2(0.260850 + 30)}2.284 = 3.061m. \tag{2}$$

Using (1.6.10), we find quite the same result,

$$\bar{y}_E = \frac{1}{\cos^2(E)}\bar{y}_T = \frac{1}{\cos^2(30)}2.284 = 3.045m.$$

c. The y-coordinate of the target on the inclined plane is

$$y_T = 503 \cdot \tan(30) = 290.41 \approx 290.5m.$$

Using RangeC.Bas:

Input: Initial x-coordinate, 0; initial y-coordinate, 0; initial speed, 883.92; departure angle, 30.26085; x-coordinate, 503; X-coordinate of another point, X = 600; BC = 3.360

Output: For x = 503 we find that the corresponding y-coordinate is 290.20, i.e. approximately equal to 290.41m.

d. Correction term,

$$\alpha_{SD} - \alpha_{SO} = \frac{h_s}{x_T} \frac{10800}{\pi} (\cos(E) - 1) = \frac{0.0381 \cdot 10800}{\pi(503)} (\cos 30 - 1) = -0.035 MOA,$$

is insignificant.

Example 1.20

A Sierra bullet caliber 0.30, 168 grain HPBT is fired with an initial speed of 807.72m/s to hit a target located at the inclined range 600meter. The atmosphere is standard and the gun is at the sea level.

Find the super elevation angle needed to hit the given target at the inclined range 600meter, if the site angle is 45 degree. The Siacci coefficient of the given Sierra bullet for the range 600m is 2.8392.

It is found that at the horizontal range 424.30m the projectile drop is 1.757m.

Find as well the drop of the given projectile at the given inclined range.

Solution

The coordinates of the target are:

$$x = D \cdot \cos(E) = 600 \cdot \cos(45) = 424.30m \,,$$

and

$$y = D \cdot \sin(E) = 600 \cdot \sin(45) = 424.30m \,.$$

For angles of departure close to zero, formula (1.5.1) shows that the drops of two respective projectiles are equal. Thus, for the angle of departure needed to hit the target located at the horizontal range 600 meters we can write:

$$\tan\alpha_0 = \frac{1.757}{424.30} = 0.004141 \,.$$

Hence, we find that

$$\alpha_0 = 0.237257° = 14.235 \ MOA \,.$$

Using (1.6.2), we write:

$$\sin(2\bar{\alpha}_0 + 45) = \sin(2 \cdot 0.237257)\cos(45) + \sin(45) = 0.710035 \,.$$

Hence, we find that the super elevation angle is

$$\bar{\alpha}_0 = 0.238248° = 14.295 \ MOA \,.$$

Using (1.5.1) or (1.6.3), we find that the drop of the bullet at the inclined range 600 meters is

$$\bar{y}_E = \frac{\cos^2(\alpha_0)}{\cos^2(\bar{\alpha}_0 + E)}\bar{y}_T = \frac{\cos^2(0.237257)}{\cos^2(0.238248 + 45)}1.757 = 3.54m \,. \qquad (1)$$

Example 1.21

A Russian bullet, caliber 7.62mm, 7.9 gr. is fired with an initial speed of 735m/s from a Simonov SKS rifle. The departure angle that corresponds to the zero range 300m is 0.216 degree. The time of flight is 0.50seconds.

Find the super elevation (as a function of the site angle E) angle needed to hit a target on the inclined plane at a point with abscissa 300m.

Apply the results for the site angles respectively equal to 30 degrees, 45 degrees, and 60 degrees.

Solution

The y-coordinate and the inclined range of the target as functions of the site angle are respectively,

$$y = x \cdot \sin(E), \quad D = x / \cos(E).$$

The drop of the projectile zeroed at the horizontal range 300 meters is

$$\bar{y}_1 = x \cdot \tan(\alpha_1) = 300 \cdot \tan(0.216) = 1.13 m.$$

The y-coordinate of the target is

$$y = x \cdot \tan(E).$$

Using (1.5.1) for the drop of the given projectile zeroed at the inclined distance,

$$D = x / \cos(E),$$

we can find the drop as a function of site angle E:

$$\bar{y}_2 = \bar{y}_1 \cdot \frac{\cos^2(\alpha_1)}{\cos^2(\bar{\alpha}_2 + E)} = 1.13 \cdot \frac{\cos^2(0.216)}{\cos^2(\bar{\alpha}_2 + E)} = \frac{1.13}{\cos^2(E)}.$$

Applying the last formula, and following the same way as in example 1.6, we find the following approximate projectile drops:

Site angle, 30 degree; super elevation angle 0.215°; inclined range, 346.40m; drop, 1.50m; Site angle, 45 degree; super elevation angle 0.215°; inclined range 424.30m; drop, 2.25m; Site angle, 60 degree; super elevation angle, 0.214⁰; inclined range, 600; drop, 4.52m.

Comment

Since for a site angle of zero degree, and range 300 meters, the departure angle is 0.216 degree, based on the results obtained above using (1.5.1), we can say that practically the super elevation angle remains the same for all site angles.

We only need to aim to direct the line of sight of the rifle at the inclined range of 300 meters, and the bullet will hit the target at the inclined range 347m.

Example 1.22

For the Russian bullet of Example 1.21 use the formula (1.6.2) to find the super elevation angle, needed to hit a target on a 30 degree inclined plane at a point with abscissa 300m (Inclined distance 346.50m).

For the rifle zeroed at the horizontal range 300m the angle of departure is 0.216 degree.

Find as well the maximum "height" of the inclined trajectory from the inclined plane, if the coordinates of the vertex of the horizontal trajectory are (183m, 0.31m).

Solution

Super elevation angle

Employing (1.6.2), for the super elevation angle, we have:

$$\sin(2\bar{\alpha}_0 + 30) = \sin(2 \cdot 0.216)\cos(30) + \sin(30) = 0.506653.$$

Hence, we find that the super elevation angle is approximately equal to the angle of departure that corresponds to the horizontal range, i.e.

$$\bar{\alpha}_0 = 0.2165^0 \approx \alpha_0.$$

Vertical "distance"

Since the inclined trajectory has not a maximum height, we can consider the "vertical height" of that point on the trajectory that has the largest distance from the inclined plane along y-axis.

The drop of the horizontal trajectory at the point of maximum height is

$$\overline{y} = x\tan(\alpha_0) - y_0 = 183\tan(0.216) - 0.31 = 0.38m.$$

Using 1.6.3, for the drop of the inclined trajectory at the point of the "maximum height", we can write:

$$\overline{y} = \frac{\cos^2(0.207)}{\cos^2(0.2074 + 30)}0.35 = 0.43m.$$

For the corresponding "vertical distance", of the projectile "h" at the "inclined trajectory" we have:

$$h = [x\tan(\overline{\alpha}_0 + 30) - x\tan(30) - \overline{y} = 183\tan(0.2074 + 30) - 183\tan(30) - 0.40 = 0.46m$$

This "vertical height" is located on the inclined plane at a distance of

$$x_M = 183 \cdot \cos(30) = 211m$$

from the muzzle of the firearm.

If we aim at the center of a target located 211 meters from the firearm muzzle the bullet will deviate vertically up 0.46 meters from the center of the target.

Those characteristics should be considered for the inclined fire.

1.7 Inclined Fire and the Rifleman's Rule 2

Consider the projectile trajectory launched with initial speed v_0, departure angle α_0, where $|\alpha_0| \leq 15°$. The impact point x_T is at the horizontal range D_0. Consider as well another trajectory of the same projectile (same ballistics coefficient, same initial speed) launched with an angle of $(\overline{\alpha}_0 + E)$, whose impact point "E" is on the inclined range at the distance " D ", (Figure 5).

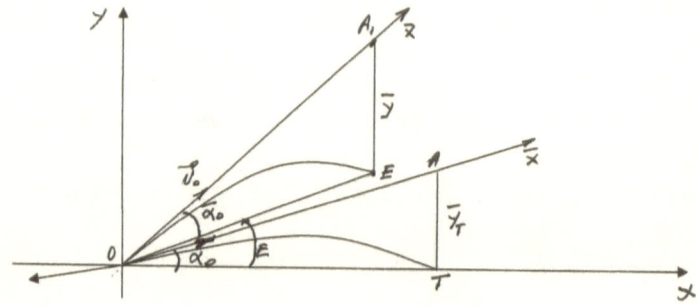

Figure 5—Inclined Shooting

$\vec{v}_0$—Departure velocity; EOA—Super Elevation Angle; TOE—Angle of Site;

O—Location of the Muzzle of the GUN.

E—Point of Impact; OE = D; OT = Do

For such trajectories, it is found (Ref. Okunev, B. H, "Fundamentals of Ballistics", page 240, Vol.1, Book 2, Moscow, 1943) that:

- The projectile drop for both trajectories is the same, i.e.

$$\bar{y} = \bar{y}_T, \tag{1.7.1}$$

if the super elevation angle and the inclined distance are respectively given by the equations:

$$\sin(\bar{\alpha}_0) = \sin(\alpha_0)\cos(E), \tag{1.7.2}$$

and

$$D = D_0 \frac{\cos(\bar{\alpha}_0 + E)}{\cos(\alpha_0) \cdot \cos(E)}, \tag{1.7.3}$$

where $\bar{\alpha}_0$ is the super elevation angle.

Using (1.7.2) and (1.7.3), we can find the super elevation angle $\bar{\alpha}_0$ and the corresponding inclined range "D" when we know the departure angle α_0 and the corresponding horizontal range D_o. Thus, we have another rule expressed by (1.7.2) and (1.7.3) to find the super elevation angle and to set up the corresponding aiming angle. We will call that rule Rifleman's Rule 2.

Equations (1.7.1), (1.7.2) and (1.7.3) can be easily obtained from Figure 5, using the sine for the triangles EOA and TOB.

For narrow departure angles, the Rifleman's Rule 2 can be further simplified. Indeed, for narrow angles we can consider that:

$$\sin(\alpha_0) \approx \alpha_0 , \text{ and } \sin(\overline{\alpha}_0) \approx \overline{\alpha}_0 , \qquad (1.7.4)$$

where α_0 and $\overline{\alpha}_0$ are expressed in radian.

From (1.7.2) we find that the super elevation angle $\overline{\alpha}_0$ is

$$\overline{\alpha}_0 \approx \alpha_0 \cdot \cos(E) , \qquad (1.7.5)$$

while from (1.7.3) we find that

$$D \approx D_0 . \qquad (1.7.6)$$

The conditions (1.7.4) are satisfied for relatively short firing ranges, as for example, for the fire within the effective range of infantry rifles, machine guns, hunting rifles, etc.

The simplified Rifleman's Rule 2, can be easily used in the practice of shooting with small arms. To find the superelevation angle needed to zero the rifle at the inclined range $D \approx D_0$ we have to calculate the super elevation angle using (1.75.) and set up the corresponding angle of sight.

Note that we have reached the same result expressed by (1.7.5) and (1.7.6) for the ideal projectile flying in absence of drag (see G. Klimi, "Exterior Ballistics with Applications", p. 51, Xlibris, 2008).

Angle of Sight

The angle of sight (in MOA) that zeroes the firearm at the inclined range $D \approx D_0$ is

$$\alpha_{SD} = 60 \cdot \alpha_0 \cos E + \frac{h_S}{D_0} \frac{10800}{\pi} , \qquad (1.7.7)$$

while the angle of sight that zeroes the firearm at the same horizontal range D_0 is

$$\alpha_{S0} = 60 \cdot \alpha_0 + \frac{h_S}{D_0} \frac{10800}{\pi} . \qquad (1.7.8)$$

The departure angle is supposed to be in degree.

From (1.7.7) and (1.7.8), we find that the "inclined" angle of sight in MOA is

$$\alpha_{SD} = \alpha_{S0} + 60 \cdot \alpha_0 (\cos E - 1) . \qquad (1.7.9)$$

To adjust the angle of sight to hit the target at the inclined range $D \approx D_0$, we need to correct the angle of sight that corresponds to the horizontal range $D \approx D_0$ adding the (negative) term

$$\alpha_{SD} - \alpha_{S0} = 60 \cdot \alpha_0 (\cos E - 1) . \qquad (1.7.10)$$

Comment

The method of finding the angle of sight by the Rifleman's Rule 2, is more practical then the rifleman's rule for use in practice of shooting since the range tables are given usually for distances 100, 200, . . ., 1000 meters. Therefore, it is easy to adjust the angle of sight for the inclined distances 100-1000 meters only by correcting the "horizontal" angle of sight adding the (negative) quantity (1.7.10).

Example 1.23

The range table of the Russian cannon 122mm gives a horizontal range of 6400m for the 122mm projectile fired with an initial speed of 885m/s at an angle 3.20 degree, in standard conditions.

If the site angle is 30 degree, find the corresponding super elevation angle and the inclined distance where the projectile will hit the target.

Use the PC program Range122.Bas to verify that the result we obtain is acceptable.

Solution

The drop of projectile at the impact point on the horizon is

$$\bar{y}_0 = D_0 \cdot \tan(E) = 6400 \cdot \tan(30) = 3695m.$$

Employing (1.7.4), we find that the super elevation angle needed to set up the "the sight mechanisms" in order to have the same drop is

$$\bar{\alpha}_0 = \sin^{-1}[\sin(\alpha_0)\cos(E)] = \sin^{-1}[(\sin 3.20)\cos(30)] = 2.77092°.$$

The inclined distance to the point of impact (see (1.7.2)) is

$$D = D_0 \frac{\cos(\bar{\alpha}_0 + E)}{\cos(\alpha_0)\cdot\cos(E)} = 6400 \cdot \frac{\cos(2.77092 + 30)}{\cos(3.20)\cdot\cos(30)} = 6223m.$$

The x-coordinate and y-coordinate of the point of impact are respectively

$$x = D \cdot \cos(E) = 6223 \cdot \cos(30) = 5389.80m,$$
and

$$y = D \cdot \sin(E) = 6223 \cdot \sin(30) = 3112m$$

Using PC Program Range122.Bas

Input: y-coordinate of cannon; departure angle, 32.77092; Departure speed, 885; x-coordinate of a point on the trajectory, 5390.
Output: y-coordinate of the point with abscissa 5390m is 3118.

Remark

Comparing the result (3112m), obtained using the rifleman's rule 2, i.e. the set of formulas (1.7.1) and (1.7.2), with the result (3118m) obtained using the PC program, we see that the accuracy obtained using the approximate formulas is acceptable.

Example 1.24

Find the super elevation angle needed to hit a target located 4200 meters from the muzzle of 122mm Russian cannon, if the site angle of the target is 25 degree. The departure speed of the projectile is 885m/s.

Use the PC program to verify the accuracy of the obtained data.

Solution

From the range table of 122mm Russian cannon we find that the departure angle needed to hit the target located at the horizontal range 4200 meters is 1.9 degree, while for 4400m is 2 degree.

Using the above data we find the super elevation angle and the inclined distance are:

For the range 4200m,

$$\bar{\alpha}_0 = \sin^{-1}[\sin(\alpha_0)\cos(E)] = \sin^{-1}[(\sin 1.9)\cos(25)] = 1.721928°,$$

and

$$D = 4200 \frac{\cos(1.721928 + 25)}{\cos(1.721928) \cdot \cos(25)} = 4141.12m.$$

For the range 4400m,

$$\bar{\alpha}_0 = \sin^{-1}[\sin(\alpha_0)\cos(E)] = \sin^{-1}[(\sin 2)\cos(25)] = 1.812550°,$$

and

$$D = 4400 \frac{\cos(1.812550 + 25)}{\cos(1.812550) \cdot \cos(25)} = 4335.07m.$$

Using the interpolation, we find that the super elevation angle is

$$\bar{\alpha}_0 = \frac{(4200 - 4141.12)}{(4335.07 - 4141.12)} \cdot (1.812550 - 1.721928) + 1.721928 = 1.749498°.$$

Thus, to hit the target located at the given inclined distance the departure angle should be approximately

$$\bar{\alpha}_0 = 1.749498°.$$

Using PC Program Range122.Bas

Input: Departure angle, 26.1749; Initial speed, 885; the x-coordinate of a point on the trajectory, equal to the projection of the inclined distance on the x-axis,

$$x = 4200 \cdot \cos(25) = 3806.$$

Output: y-coordinate 1778.50m; impact angle, -22.91225 degree.

Remark

The y-coordinate of the point of impact on the inclined distance is

$$x = 4200 \cdot \sin(25) = 1775m.$$

The projectile will pass 3.5 meters over the target in y-direction, i.e. it will explode on the inclined plane about 4.5m far from the target.

Example 1.25

A Sierra bullet caliber 0.30, 168 grain HPBT is fired with an initial speed of 807.72m/s to hit a target located at the inclined range 600 meters. The atmosphere is standard and the shooting takes place at the sea level.

a. Find the super elevation angle needed to hit the given target at the inclined range 600meter, if the site angle is 45 degree.
   For the departure angle $\alpha_0 = 0°$, the drop of the projectile at 600m is 3.806m.
b. Use RangeC.Bas to verify the results. The Siacci coefficient of the given Sierra bullet is 2.8554.

Solution

a. Super Elevation Angle

The drop of projectile zeroed at 600 meters is also 3.806m. It can easily be seen that the bullet departure angle, α_0, needed to hit the target at the horizontal range 600 is

$$\alpha_0 = \tan^{-1}(\frac{3.806}{600}) = 0.363441° = 21.806 MOA.$$

The drop of the projectile at the inclined distance 600 meters is also 3.806m (see 1.7.1), but the departure angle $\bar{\alpha}_0$ need to be estimated. Using (1.7.5), we find that the super elevation angle is

$$\bar{\alpha}_0 \approx \alpha_0 \cdot \cos(E) = 0.363441 \cdot \cos(45) = 0.256992° = 15.42 MOA.$$

The inclined distance is

$$D = D_0 \frac{\cos(\bar{\alpha}_0 + E)}{\cos(\alpha_0) \cdot \cos(E)} = 600 \cdot \frac{\cos(0.256992 + 45)}{\cos(0.363441) \cdot \cos(45)} = 597.31m.$$

Using the estimated angle we find a smaller value of the inclined distance (597.31) that is around 2.68 meters less than 600m.

We can accept the estimated angle and consider the inclined range 600 meters, since the deviation of the bullet in the vertical plane is small, around

$$2.68 \cdot \tan(0.256992) = 0.012m = 1.2cm.$$

Remark

To hit the target located at the inclined distance 600m, we need to use the estimated super elevation angle $\bar{\alpha}_0 = 15.42 \; MOA$. This angle is around 6.40 MOA smaller that the corresponding horizontal angle $\alpha_0 = 21.806 \; MOA$.

If we do not adjust the super elevation angle, and related with that the corresponding angle of sight, i.e. if we use the value $\alpha_0 = 0.363481°$ that corresponds to the horizontal range 600m, the bullet will deviate around one meter from the center of the target.

Adjusting the angle of sight

Method I

The angle of sight that zeroes the firearm at the same horizontal range $D_0 = 600m$ is

$$\alpha_{S0} = 60 \cdot \alpha_0 + \frac{h_S}{D_0} \frac{10800}{\pi} = 60 \cdot (15.42) + \frac{(0.0381) \cdot 10800}{600\pi} = 22.024 MOA.$$

The "correction" term (1.7.10) is

$$\alpha_{SD} - \alpha_{S0} = 60 \cdot \alpha_0 (\cos E - 1) = 21.806 \cdot [(\cos 45) - 1] = -6.387 MOA.$$

Thus, we find that the angle of sight needed to zero the firearm at the inclined range $D_0 = 600m$ is

$$\alpha_{SD} = \alpha_{S0} + \alpha_0 (\cos E - 1) = 22.024 - 6.387 = 15.637 MOA.$$

Note

To fire at the inclined range 600 meters we need to correct the horizontal sight angle subtracting $6.387 MOA$ from the sight angle $\alpha_{S0} = 22.024 MOA$.

Method II

Using (1.7.7), we find that angle of sight (in MOA) that zeroes the firearm at the inclined range $D \approx 600m$ is

$$\alpha_{SD} = 60 \cdot \alpha_0 \cos E + \frac{h_S}{D_0} \frac{10800}{\pi} = 21.806 \cos(45) + \frac{(0.0381) \cdot 10800}{600\pi} = 15.637 MOA \cdot$$

b. Using RangeC.Bas

Input: Departure angle, 45.256992; Initial speed 807.72; BC, 2.8554; x-coordinate of a point, 424 (i.e. the x coordinate of the target that is in the inclined distance 600m).
Output: y-coordinate (that corresponds to x-coordinate 424) is 423.80.

The result obtained shows that the projectile will hit the target at the inclined distance 600m, located at the point with coordinates (424m, 424m).

1.8 Again on the Inclined Fire

Projectile Drop

We will show another approach to estimate the super elevation angle, for relatively short inclined distances.
For short inclined distances, the equation of projectile trajectory in curvilinear coordinates (Ref. Klimi, G. "Exterior Ballistics with Applications", p. 161, Xlibris 2008) is given by:

$$\bar{y} = \frac{g}{2 \cdot u_0^2} \cdot \bar{x}^2 + \frac{gB(u_0 - 240)}{3u_0^4} \bar{x}^3 + gB^2(u_0 - 240)\frac{u_0 - 320}{4u_0^6 \cdot} \bar{x}^4, \quad (1.8.1)$$

or, by the less accurate trajectory equation,

$$\bar{y} = \frac{g}{2 \cdot u_0^2} \cdot \bar{x}^2 + \frac{gB(u_0 - 240)}{3u_0^4} \bar{x}^3, \quad (1.8.2)$$

where $B = ch(y)/3$, E is the site angle, $\bar{\alpha}_0$ is the super elevation angle, $u_0 \approx v_0$ is the departure speed while u is the component of projectile velocity $\bar{v}$ along $\bar{x}$-axis (Figure 2).

For narrow super-elevation angle (angle of sight), we can assume that $u \approx v$. In equation (1.8.1), the quantity $\bar{y}$ is the drop of the projectile that corresponds to a given value of the curvilinear coordinate $\bar{x}$.

The trajectory equation of the projectile flight (1.8.1), or (1.8.2), shows that:

- All projectiles with the same ballistics coefficient (i.e. same $B = ch(y)/3$) that are launched with the same initial speed v_0 for the same curvilinear coordinate $\bar{x}_E$ have the same drop $\bar{y}$, regardless of the departure angle.

In other words, the families of projectile trajectories with the same ballistics coefficient and same initial speed have the same drop $\bar{y}$ for the same value of curvilinear coordinate $\bar{x}$.

Using the theorem of sine respectively for the triangles in Figure 4, we find that

$$x_E = x_0 \cdot \frac{\cos(\bar{\alpha}_0 + E)}{\cos(\alpha_0)}. \qquad (1.8.3)$$

From the above equations, it can be easily seen that the drop of the projectile launched with a super elevation angle $\bar{\alpha}_0$ (departure angle $\bar{\alpha}_0 + E$) is equal to the drop of the same projectile launched with an angle α_0, if the distances of the respective points are related with the following equations

$$D_E = D_0 \cdot \frac{\cos(\bar{\alpha}_0 + E)}{\cos(\alpha_0) \cdot \cos(E)}.$$

Thus, we can estimate the drop of a projectile in an inclined fire if we know the drop of the same projectile fired on the horizontal plane with departure angle α_0.

Super Elevation Angle

The result we obtained above for the inclined fire allows us to find the super elevation angle $\bar{\alpha}_0$ needed to hit the target located at a given slant range D.

From Figure 4, we can express the curvilinear coordinates of the target $(\bar{x}_E, \bar{y}_E)$ through the Cartesian coordinates of the target (x_E, y_E),

$$\bar{x}_E = x_E / \cos(\bar{\alpha}_0 + E) \qquad\qquad (1.8.4) \text{ and}$$

$$\bar{y}_E = x_E \tan(\bar{\alpha}_0 + E) - x_E \tan(E) \qquad\qquad (1.8.5)$$

Substituting (1.8.4) and (1.8.5) into (1.8.2) we obtain the trajectory equation in Cartesian coordinates:

$$x_E \tan(\bar{\alpha}_0 + E) - x_E \tan(E) = \frac{g}{2 \cdot u_0^2} \cdot \frac{x_E^2}{\cos^2(\bar{\alpha}_0 + E)} + \frac{gB(u_0 - 240)}{3u_0^4} \cdot \frac{x_E^3}{\cos^3(\bar{\alpha}_0 + E)}$$

$$(1.8.6)$$

The left side of the equation (1.8.6) is the projectile drop.

When the target is in the same level with the gun, the site angle is zero. Using (1.8.6) for the projectile drop when site angle E is zero, while the departure angle is α_0 we can write

$$x_0 \tan(\alpha_0) = \frac{g}{2 \cdot u_0^2} \cdot \frac{x_0^2}{\cos^2(\alpha_0)} + \frac{gB(u_0 - 240)}{3u_0^4} \cdot \frac{x_0^3}{\cos^3(\alpha_0)}, \qquad (1.8.7)$$

since the y-coordinate of the impact point is zero.

From (1.8.6) and (1.8.7), it follows that the projectile drop will be the same, i.e.

$$x_E \tan(\bar{\alpha}_0 + E) - x_E \tan(E) = x_0 \tan\alpha_0, \qquad\qquad (1.8.8)$$

if

$$\frac{x_E}{\cos(\bar{\alpha}_0 + E)} = \frac{x_0}{\cos(\alpha_0)}. \qquad\qquad (1.8.9)$$

Employing (1.8.9), equation (1.8.8) can be written

$$\sin(\bar{\alpha}_0 + E) \cdot \cos(E) - \sin(E) \cdot \cos(\bar{\alpha}_0 + E) = \sin\alpha_0 \cdot \cos(E). \qquad (1.8.10)$$

Hence,

$$\sin(\bar{\alpha}_0) = \sin\alpha_0 \cdot \cos(E) \qquad (1.8.11)$$

when, see (1.8.9),

$$x_E = x_0 \frac{\cos(\bar{\alpha}_0 + E)}{\cos(\alpha_0)}, \text{ or } D_E = D_0 \cdot \frac{\cos(\bar{\alpha}_0 + E)}{\cos(\alpha_0) \cdot \cos(E)}. \qquad (1.8.12)$$

For small arms and relatively short distances, the departure angle and the super elevation angles are relatively narrow, so we can assume that

$$\text{X } \cos(\bar{\alpha}_0 + E) = \cos(E), \ \cos(\alpha_0) = 1. \qquad (1.8.13)$$

Assuming (1.8.13) is true, we find that

$$\sin(\bar{\alpha}_0) = \sin\alpha_0 \cdot \cos(E) \qquad (1.8.14)$$

when

$$x_E = x_0 \cos(E), \text{ or } D_E \approx D_0. \qquad (1.8.15)$$

For small values of the departure angle (1.8.14) can be written:

$$\bar{\alpha}_0 = \alpha_0 \cdot \cos(E). \qquad (1.8.16)$$

Formula (1.8.11) and formula (1.8.16) show that the super elevation angle $\bar{\alpha}_0$ depends on the site angle E; for the same inclined distance, the super elevation angle decreases as the site angle increases.

The results obtained for the inclined fire allow us to find and set up the aiming angle (super elevation angle $\bar{\alpha}_0$), and to calculate the projectile drop when we know the drop of the projectile and the departure angle α_0 that zeros the trajectory at the horizontal range D.

They are valid for all projectiles that have the same ballistics coefficient, and are launched with the same initial speed.

The following examples illustrate the use of the results obtained in the last two sections

1.9 Siacci's Trajectories

The projectile trajectory for the flight of the high-speed projectiles for departure angles -15 degrees to 15 degrees is described by the equations we have obtained in section 5.2 of "Exterior Ballistics with Applications":

$$x = -\frac{1}{Bg}[D(u) - D(v_0)], \tag{1.9.1}$$

$$y = p_0 x + \frac{x}{240 B \cos^2 \alpha_0} \cdot [\frac{A(u) - A(v_0)}{D(u) - D(v_0)} - J(v_0)], \tag{1.9.2}$$

$$t = -\frac{1}{Bg \cos \alpha_0}[T(u) - T(v_0)], \tag{1.9.3}$$

$$p = p_0 + \frac{1}{240 B \cos^2 \alpha_0}[J(u) - J(v_0)], \tag{1.9.4}$$

where, for the trajectories of the projectiles fired to hit the target located at the same horizontal plane,

$$B = \frac{c}{3g\sqrt{\cos(\alpha_0)}} h(\overline{y}). \tag{1.9.5}$$

For departure angles α_0 in the interval -15 degree to 15 degree, the equations (1.9.1)-(1.9.4) show that at the horizontal range on the ground, the elements of the trajectory depend on the ballistics coefficient c, and the departure angle α_0 (we assume a constant value of the density function $h(\bar{y})$)). Thus, all the elements of the trajectory of a projectile fired with an initial speed v_0 are determined by the departure angle α_0 and the ballistics coefficient c (or equivalently B).

Siacci's Trajectories of Different Projectiles

Consider two projectiles respectively with ballistics coefficients c_1 and c_2 (corresponding parameters B_1 and B_2) launched with the same initial speed v_0, but with different departure angles, respectively α_{01} and α_{02}.

Consider the elements of the trajectory at the impact point of the first projectile, i.e. respectively the projectile range, y-coordinate, time, terminal speed, terminal pseudo speed, and terminal angle:

$$x_{1T}, \ y_{1T} = 0, \ t_{1T}, \ v_{1T}, \ u_{1T} \ \text{and} \ \alpha_{1T} \ (\text{corresponds to } p_{1T}).$$

Employing (1.9.1)-(1.9.4) for each projectile, we obtain the following formulae that show the relation of the elements of the second trajectory with the elements of the point of impact of the first projectile (for the same value of the pseudo speed u_{1T}):

$$x_2 = \frac{B_1}{B_2} x_{1T}, \tag{1.9.6}$$

$$y_2 = x_2 (p_{02} - \frac{B_1}{B_2} \frac{\cos^2 \alpha_{01}}{\cos^2 \alpha_{02}} p_{01}), \tag{1.9.7}$$

$$t_2 = \frac{B_1}{B_2} \frac{\cos(\alpha_{01})}{\cos(\alpha_{02})} t_{1T}, \tag{1.9.8}$$

$$p_2 = p_{02} - \frac{B_1}{B_2} \frac{\cos^2(\alpha_{01})}{\cos^2(\alpha_{02})} (p_{01} - p_{1T}). \tag{1.9.9}$$

Using (1.9.6)-(1.9.9), we can find the elements of the second projectile trajectory when we know the elements of the first trajectory at the point of impact, on condition that both projectiles have the same initial speed.

Siacci's Trajectories of the Same Projectile

We can use (1.9.6)-(1.9.9) for the same projectile fired with different angles.

Indeed, employing (1.9.6)-(1.9.9) for the same bullet within the effective range of the gun we can consider $c_1 = c_2$, and $h(\bar{y}_2) \approx h(\bar{y}_1) \approx 1$. Thus, we can write the above formulas in the following form:

$$x_2 = x_{1T}, \qquad (1.9.10)$$

$$y_2 = x_{1T}\left(p_{02} - \frac{\cos^2 \alpha_{01}}{\cos^2 \alpha_{02}} p_{01}\right), \qquad (1.9.11)$$

$$t_2 = \frac{\cos(\alpha_{01})}{\cos(\alpha_{02})} t_{1T} \qquad (1.9.12)$$

$$p_2 = p_{02} - \frac{\cos^2(\alpha_{01})}{\cos^2(\alpha_{02})}(p_{01} - p_{01T}) \qquad (1.9.13)$$

For relatively narrow angles that are within the effective range of small arms (till around 800-1000 meters), when we can consider that

$$\cos^2(\alpha_{02}) \approx \cos^2(\alpha_{01}) \approx 1,$$

the formulae (1.9.11) and (1.9.13) can be written respectively as:

$$y_2 = x_{1T}(p_{02} - p_{01}), \qquad (1.9.14)$$

and

$$p_2 = p_{02} - (p_{01} - p_{01T}). \qquad (1.9.15)$$

The formula (1.9.14), after some simple transformations, can be written as

$$y_2 = x_{1T}\sin(\alpha_{02} - \alpha_{01}), \qquad (1.9.16)$$

or simply

$$y_2 = x_{1T}(\alpha_{02} - \alpha_{01}). \qquad (1.9.17)$$

In (1.9.17), α_{01} and α_{02} are expressed in radian.

Formula (1.9.17) can be further simplified for narrow departure angles if we involve the projectile drop respectively at two "zeroing" ranges, i.e.

$$p_{02} = \tan\alpha_{02} = \frac{\bar{y}_{2T}}{x_{2T}}, \quad p_{01} = \tan\alpha_{01} = \frac{\bar{y}_{1T}}{x_{1T}}. \qquad (1.9.18)$$

Substituting (1.9.18) in (1.9.14) we find that ordinate of the second bullet at the corresponding trajectory point with abscissa x_{1T} :

$$y_2 = x_{1T}(\frac{\bar{y}_{02}}{x_{2T}} - \frac{\bar{y}_{01}}{x_{1T}}). \qquad (1.9.19)$$

The Bullet Drop

For the drop of a bullet fired with a narrow angle α_0 at a given point on the trajectory (a given x-coordinate) we can write

$$\bar{y} = x \tan(\alpha_0) \qquad (1.9.20)$$

or,

$$\bar{y} = x \cdot \alpha_0, \qquad (1.9.21)$$

where the angle α_0 in (1.9.21) is in radian.

If the angle is expressed in degree then we have to transform it in radian multiplying the results by the quotient ($\pi/180$).

Note. If we know the drop of a bullet under the horizontal line, we find the departure angle needed to zero the gun at the horizontal range x, using the formulas:

$$\tan(\alpha_0) = \frac{\overline{y}}{x}, \text{ or } \alpha_0 = \frac{\overline{y}}{x}, \tag{1.9.22}$$

Comment

Formula (1.9.17) shows that the drop of a given bullet at a given x-coordinate does not depend by the departure angle. In other words, (1.9.17) is another expression of (1.5.2) that shows that within the effective horizontal range of the rifles, machine guns, hunting rifles, etc. the drop of a bullet for a given x-coordinate remains constant.

Example 1.26

A Russian 7.62mm bullet (Mod. 1943) is fired with a speed 735m/s at an angle 0.318 degree. The bullet hits the center of a target located at a horizontal range of 400 meters. The ballistics coefficient is 4.132.

Find the y-coordinate of the trajectory and the drop of the same bullet at the point with abscissa 400 meters from the gun if:

(a) The bullet is fired at an angle of 0.432 degree.
(b) The bullet is fired at an angle of 0.216.
(c) The bullet is fired at an angle of 5 degree.
(d) The bullet is fired at an angle of 10 degree.
(e) The bullet is fired at an angle of 15 degree.

Solution

(a) Since the ballistics coefficient is the same, we have

$$x_2 = x_{1T} = 400m.$$

We can write:

$$p_{01} = \tan(\alpha_{01}) = \tan(0.318), \quad p_{02} = \tan(\alpha_{02}) = \tan(0.432).$$

Substituting in (1.9.11), we find that the y-coordinate of the point with abscissa 400 is

$$y_2 = x_2(p_{02} - \frac{\cos^2 \alpha_{01}}{\cos^2 \alpha_{02}} p_{01}) = 400[\tan(0.432) - \frac{\cos^2(0.318)}{\cos^2(0.432)}\tan(0.318)] = 0.796m .$$

The drop of the bullet at the point with abscissa 400 is

$$\bar{y}_2 = 400 \cdot \tan(0.432) - 0.796 = 2.22m .$$

We find the same result if we use the (1.9.21). Indeed, substituting we have

$$\bar{y}_2 = x_{1T} \cdot \alpha_{01} = 400 \cdot (0.318 \cdot \frac{\pi}{180}) = 2.22m$$

(b) In the same way, we find respectively the y-coordinate and the drop of the bullet at 400:

$$y_2 = x_2(p_{02} - \frac{\cos^2 \alpha_{01}}{\cos^2 \alpha_{02}} p_{01}) = 400[\tan(0.216) - \frac{\cos^2(0.318)}{\cos^2(0.216)}\tan(0.318)] = -0.712m ,$$

and

$$\bar{y}_2 = 400 \tan(0.216) - (-0.712) = 2.22m .$$

(c) Like in (b) we find respectively the y-coordinate and the drop of the bullet at 400m:

$$y_2 = x_2(p_{02} - \frac{\cos^2 \alpha_{01}}{\cos^2 \alpha_{02}} p_{01}) = 400 \cdot (\tan(5) - \frac{\cos^2(0.318)}{\cos^2(5)}\tan(0.318)) = 32.758m ,$$

and

$$\bar{y}_2 = 400 \tan(5) - (32.758) = 2.24m .$$

The inclined range is

$$D_2 = \frac{400}{\cos(5)} = 401.50m.$$

(d) For 10 degree, we get respectively

$$y_2 = 68.24m \text{ and } \bar{y}_2 = 2.29m$$

and

$$D_2 = \frac{400}{\cos(10)} = 406m$$

(f) For 15 degree, we have:

$$y_2 = 104.80m, \ \bar{y}_2 = 2.38m.$$

The inclined range is

$$D_2 = \frac{400}{\cos(15)} = 414m.$$

Remark

As we see, for the site angles until around 15 degree, the drop of the bullet does not significantly change with the site angle. Moreover, the "inclined" ranges are almost the same.

That is the reason it is recommended that for relatively small inclined ranges we can use the same sight that is set for shooting in the horizontal range.

For angle of site 15 degree, the increase in the inclined range is relatively small (only 14m) when the angle of departure increases from 0.318 degree to 15 degree. A change of 14 meters in the range deviates the bullet vertically around

$$0.14 \cdot \tan(-15) = -0.04 m \,.$$

(In the above estimation we have considered approximately that the impact angle, in absolute value, is equal to the departure angle).

It is not worth to correct the sight angle for such a small quantity, especially when the arm has an iron sight, when we are not able to estimate the range accurately, or when the shooting has to be done instantly.

Example 1.27 Changing the Zeroing

A shooter with a Russian SKS rifle that fires a 7.62mm bullet with departure speed 735m/s wants to know the point where the bullet will hit a table located at a distance of 100 meters from the gun in order that the firearm should be zeroed at 500 meters.
From the range tables of the rifle he finds that the drop of the bullet at 100m and 500m is respectively 0.10m and 3.76m.

Solution

Using (1.9.22), we find:
The departure angles for 100 and 500 meters are respectively

$$\alpha_{01} = \tan^{-1}(\frac{0.10}{100}) = 0.0573° = 3.438 \ MOA \,,$$

and

$$\alpha_{02} = \tan^{-1}(\frac{3.77}{500}) = 0.432° = 25.92 \ MOA \,.$$

Substituting in (1.9.11), we find that the average vertical deviation of the bullets from the center of the target is

$$y_2 = 100((\tan 0.432) - \frac{\cos^2(0.0573)}{\cos^2(0.432)} \tan(0.0573)) = 0.654 m \,,$$

i.e. about 65.4 centimeters above the center of the target.

We obtain the same result if we use the formula (1.9.17):

$$y_2 = x_{1T}(\alpha_{02} - \alpha_{01}) = 100(0.432 - 0.0573) \cdot \frac{\pi}{180} = 0.654.$$

Note. The above procedure can be simplified further using the formula (1.9.19):

$$y_2 = x_{1T}(\frac{\overline{y}_{02}}{x_{2T}} - \frac{\overline{y}_{01}}{x_{1T}}) = 100(\frac{3.77}{500} - \frac{0.10}{100}) = 0.654m.$$

Example 1.28

A marksman fires a 0.30 ball M2 bullet with an initial speed of 2800 ft/s.

Where does the bullet hit a firing table located at a distance of 100 yards from the gun in order that the firearm should be zeroed at 500 yards?

The drop of the bullet at the horizontal range 100 yard and 500 yard from the gun is respectively 2.8 in. and 94.8 in.

Show as well, the aiming angle needed to hit the center of the firing table at the desired point, so that the gun would be zeroed at 500 yards. The sight height is 1.5 inches.

Solution

Substituting in (1.9.19), we find that the trajectory height is

$$y_2 = x_{1T}(\frac{\overline{y}_{02}}{x_{2T}} - \frac{\overline{y}_{01}}{x_{1T}}) = 100(\frac{76.5}{500} - \frac{2.30}{100}) = 13 \ inches.$$

To find the angle of sight we use the formula (1.7.7), i.e.

$$\alpha_{SO} = 60 \cdot \alpha_0 + \frac{h_S}{D_0} \frac{10800}{\pi}. \tag{1}$$

The departure angle needed to zero the gun at 500 yard is

$$\alpha_0 = \frac{drop}{range} = \frac{76.5}{500 \cdot 36} = 0.00425 \; radian = 2435°.$$

Substituting in (1) we find that the angle of sight that zero the gun at 500 yards, and let the bullet pass 13 inches over the center of target located at 100 yards is

$$\alpha_{SO} = 60 \cdot \alpha_0 + \frac{h_S}{D_0} \frac{10800}{\pi} = 60 \cdot (0.2435) + \frac{(13) \cdot 10800}{(500 \cdot 36)\pi} = 2.483 \; MOA.$$

With the angle of sight set up around 2.5 MOA, the shooter should aim at the point that is

$$y_s = 1.5 \frac{400}{500} = 1.2 \; inches$$

over the center of target to have the bullet hit 13 inches over the center of the firing table.

Example 1.29 Testing the Sight

In practice of firing with small arms, it is necessary to test the sight of the firearm to assure the accuracy. The firing tests can be performed for example after repairing the sight, or parts of the rifle, etc.

A shooter with a Russian rifle Simonov SKS fires some 7.62mm bullets with a departure speed of 735m/s on a firing table located 100 meters from the rifle. The angle of sight is set up to zero the rifle at the horizontal range 300 meters.

The center of the group of shots is 0.308 meters above the center of the table (or above the so called "control point").

Using the Range Table of the rifle the shooter finds that the departure angle that zeroes the rifle at 300 meters is 0.216 degree, while the departure angle that zeroes the rifle at 100 meters is 0.057 degree.

a. Is the sight of the rifle set up correctly to be zeroed at 300 meters?
b. What are the departure angle and the correction in the angle of sight that is needed to zero the rifle at 300 meters, if the sight height is 0.0381 meters?
c. How high does the rifle shoots?

Solution

a. Substituting in (1.9.17), we find that the y-coordinate of the center of the group of shots should be

$$y_2 = x_{1T}(\alpha_{02} - \alpha_{01}) = 100(0.216 - 0.057)\frac{\pi}{180} = 0.278m$$

over the center of the target and not 0.308m. That means that the rifle fires higher then is normally accepted. The sight is not set up correctly.

b. To correct the sight we estimate the actual departure angle using (1.9.17). Solving (1.9.17) for α_{02} and substituting we find that the actual angle of departure is

$$\alpha_{02} = \alpha_{01} + \frac{y_2}{x_{1T}} = 0.057\frac{\pi}{180} + \frac{0.308}{100} = 0.00407 = 0.2335°.$$

The angle of sight that is set up to zero the rifle at 300 meters must be corrected with the quantity that corresponds to

$$\Delta\alpha = (0.216 - 0.2335)\cdot 60 = -1.05 \; MOA.$$

Thus, we have to reduce the height of the scope by 1 MOA.

c. To find how high the rifle shots over the center of the target located at 300 meters we use again (1.9.17). We find that the rifle shoots

$$y_2 = x_{1T}(\alpha_{02} - \alpha_{01}) = 300(0.2335 - 0.216)\frac{\pi}{180} = 0.092m$$

over the center of the target.

We obtain the same result using the correction quantity $\Delta\alpha = -1.05 \ MOA$.

$$y_2 = x \cdot \Delta\alpha = 300(1.05\frac{\pi}{10800}) = 0.92m.$$

1.10 Experimental Determination of the Departure Angle

The departure angle can be measured in practice of shooting using a target board that is in a distance of 15-25 (or 15-20) meters from the firearm.

Firing some bullets (with a given sight zeroed at a certain range) on the target board, for example located 20 meters far from the muzzle, we determine the center of the point of impact on the table, and measure the height y_T of that center with respect to the height of the firing muzzle y_0.

The angle of departure can be estimated using the formula

$$\tan\alpha_0 = \frac{y_T - y_0 + \overline{y}_T}{x_T}, \tag{1.10.1}$$

where $\overline{y}_T$ is the drop of the projectile at the point with coordinate x_T.

Since the range of fire is short, we can use the equation of the projectile trajectory (3.3.5), (ref. Exterior Ballistics with Applications, p. 162, Xlibris, 2008) that can be written in Cartesian coordinates in the following form:

$$\bar{y}_T = x_T \tan \alpha_0 - y_T = \frac{g}{2 \cdot v_0^2 \cos^2 \alpha_0} \cdot x_T^2 + \frac{gB(v_0 - 240)}{3 v_0^4 \cos^3 \alpha_0} x_T^3. \qquad (1.10.2)$$

The left side of (1.10.2) is the projectile drop.

For short ranges, we can ignore the second term on the right side of (1.10.2) and write:

$$x_T \tan \alpha_0 - y_T = \frac{g}{2 \cdot v_0^2 \cos^2 \alpha_0} \cdot x_T^2. \qquad (1.10.3)$$

For example, for the 7.62mm projectile of the Russian rifle,

$$(B = c / (3) = (4.5663) / 3 = 1.5221),$$

we find that the third term in (1.10.2) at $x = 20m$ is very small (about $6.75 \cdot 10^{-5}$) compared to the second term (about 0.00567).

Substituting (1.10.3) into (1.10.1) we obtain the following equation:

$$\tan \alpha_0 = \frac{y_T - y_0}{x_T} + \frac{gx_T}{2 \cdot v_0^2 \cos^2 \alpha_0} \qquad (1.10.4)$$

Since the departure angle is very narrow, we can consider the square of cosine approximately one. Then, the departure angle can be obtained as the solution of the equation:

$$\tan \alpha_0 = \frac{y_T - y_0}{x_T} + \frac{gx_T}{2 \cdot v_0^2} \qquad (1.10.5)$$

For shooting with small arms, the departure angle or the aiming angle is very narrow and needs to be determined and set up with a great accuracy since a small change in the departure angle introduces a relatively great error in the firing range and in the ballistics coefficient.

The error in BC can be easily estimated taking in consideration equations (6.1.4) and (6.1.5) shown in Exterior ballistics with Applications, i.e.

$$\frac{\partial x}{\partial c} = -(1 - \frac{\tan \alpha_0}{\tan |\alpha_T|}) \frac{x}{c},$$

(1.10.6)

$$\frac{\partial x}{\partial \alpha_0} = \frac{\cos(2\alpha_0)}{\cos^2 \alpha_0 \tan |\alpha_T|} x.$$

(1.10.7)

Eliminating $\partial x / x$ from the above equations we can write:

$$\frac{\partial c}{c} = \frac{\cos(2\alpha_0)}{\cos^2 \alpha_0 (\tan |\alpha_T| - \tan \alpha_0)} \partial \alpha_0.$$

(1.10.8)

Thus, to have an idea about the error in the estimation of the BC, let's find the error we make when we determine the departure angle (or set up the sight) for the 7.62 mm Russian bullet with an error of one minute. The bullet is fired with an initial speed 735m/s, while the departure angle and the impact angle are respectively 0.432 degree and 0.732 degree.

Substituting in (1.10.8), we find that an error in the departure angle of one minute introduces a relative error in the BC of

$$\partial c / c = 0.0555 = 5.55\%$$

The BC of the given bullet estimated in example 1.12 is 4.064. Because of the error in the departure angle, the measured BC will be 4.290.

Example 1.30

The firing board is located 25 meters from the rifle.

Find the angle of departure if a bullet fired with initial speed 807.72m/s hits the firing board 8 centimeters above the center of the target. The target and the gun are at the same horizontal line. Consider a standard atmosphere, and that the tests are at the sea level.

Solution

Substituting in (1.10.5), we have:

$$\tan\alpha_0 = \frac{y_T - y_0}{x_T} + \frac{g}{2 \cdot v_0^2} \cdot x_T = \frac{0.08}{25} + \frac{9.80665 \cdot (25)}{2(807.72)^2} = 0.0033879 \,.$$

Hence, we find that the departure angle is

$$\alpha_0 = \tan^{-1}(0.0033879) = 0.194111° \,.$$

1.11 Range Tables of Small Arms

The range table of a given small firearm includes data needed to have a practical, easy and at the same time accurate fire. It contains the data of the bullet trajectory for some given ranges, usually for distances within the effective range of a firearm, for example 100m, 200m, 300m ... 800m and for shooting in the standard atmosphere.

Some range tables of small arms contains as well the "corrections" for the changes in range as result of small deviations of the temperature and pressure of air from standard atmosphere, and the initial speed, the temperature of black powder, etc., from the standard values, as well as the corrections of the range as result of the range-wind and crosswind.

Since for small changes of the temperature, pressure, etc. the change in range is relatively small, it results that the changes in vertical direction are insignificant.

The main factor that influence the accuracy of shooting is the cross wind that deflects the bullet from the plane of shooting to the left or to the right from.

The compilation of range tables is already studied in the "Exterior Ballistics with Applications". There are displayed many examples and the use of the PC programs to construct the Range and Ballistics Tables.

For example, the corrections or small arms are illustrated in example 1, page 306 of the "Exterior Ballistics with Applications.

Hereafter (table 1) is given the Range Table of 0.30 M2 Ball bullet, fired with an initial speed 853.44m/s (2800ft/s) in standard atmosphere.

The calculations are performed using the PC program AngleM30. Bas presented hereafter in the book. The range table can also be obtained using the PC program AngleC.Bas but using a BC = 3.248.

Table 1—Caliber 0.30 Ball M2, Initial Speed 853.44m/s. Standard Atmosphere

Range m	Departure Angle (degree)	Time s	Speed m/s	Terminal Angle degree	Cross wind 5m/s Deflection m	Trajectory Vertex m
100	0.0410	0.123	777	-0.0438	0.028	(51, 0.018)
200	0.0878	0.258	704	-0.0997	0.118	(103, 0.082)
300	0.1413	0.408	635	-0.1721	0.281	(157, 0.204)
400	0.2030	0.574	569.5	-0.2600	0.527	(214, 0.405)
500	0.2750	0.76	509.5	-0.3882	0.869	(272, 0.710)
600	0.3592	0.97	455	-0.5468	1.322	(332, 1.150)

The data that are in table 1 are approximately equal to the respective data obtained using Robert McCoy's PC program MCTRAJ41.EXE, for the same projectile.

Example 1.31

Use the data given in the range table1 to find the angle of sight needed to zero the gun at 500 meters if the sight height is 5cm, as well as the drop of the bullet at 500 meters.

Solution

The departure angle that zeroes the bullet at 500 meters is

$$\alpha_0 = 0.275° = 16.5 MOA.$$

Using the formula (1.7.7), we find that the sight angle is

$$\alpha_{S0} = 60 \cdot \alpha_0 + \frac{h_S}{D_0} \frac{10800}{\pi} = 60 \cdot (0.275) + \frac{(0.05) \cdot 10800}{600\pi} = 16.786 MOA.$$

The drop of projectile at 500 meters is

$$\bar{y} = x \tan \alpha_0 = 500 \tan(0.275) = 2.40 m$$

under the horizontal line.

1.12 Small Arms Firing on the Mountains

In high mountains, there are big changes in temperature, pressure and density of air with respect to the normal standard values that are present at the sea level. The projectile moves in a less dense atmosphere where the pressure and the temperature of air are considerably less than the respective values at the sea level.

Standard Atmosphere

For the standard atmosphere, the virtual temperature, the density and the pressure of air at a given altitude can be estimated respectively using the formulas (ref. section 8.2, "Exterior ballistics with Applications"):

$$\tau = 289.08 - 0.006328 \cdot y, \qquad\qquad (1.12.1)$$

$$\rho = 1.205 \cdot (\frac{289.08 - 0.006328 \cdot y}{289.08})^{4.4}, \qquad\qquad (1.12.2)$$

$$p = 750 \cdot (\frac{289.08 - 0.006328 \cdot y}{289.08})^{5.4}. \qquad\qquad (1.12.3)$$

The virtual temperature can be estimated using formula (8.1.10), page 378, ref. Exterior Ballistics wit Applications,

$$\tau = \frac{T}{1 - 0.3785 \cdot (e/p)}, \qquad (1.12.4)$$

where "e" is the pressure of saturated water vapor (pressure w.v) that can be estimated using the data presented in the following table (temperature in degree Celsius, pressure in mm. Hg).

Table 1—Saturated Pressure of Water Vapor and pwv (humidity 100%).

Temp.	-40	-18	-10	0.0	5.0	10	15	20	25	30	38	40	54
P. w.v	0.15	1.14	1.95	4.58	6.54	49.20	12.7	17.54	23.76	31.7	49.2	55.1	115.1

Note. If the humidity is less than 100% (for example it is 50%), the p.w.v. that corresponds to humidity 50% is obtained by multiplying the respective value in table 1 with 0.50. Thus, at 15 degree Celsius (288.15 Kelvin) and 50% humidity, the pressure of w.v. is 6.35mm Hg. Substituting the above value in (1.12.4) we find that the virtual temperature is 289.08 Kelvin.

If the atmosphere at the sea level is not standard, then to estimate the temperature, the density, and the pressure of air at a given altitude, we can use again the formulas (1.12.1)-(1.12.3), substituting instead of 289.08 the value of the virtual temperature at the sea level.

If we measure the temperature at a certain altitude, as well as the humidity and the pressure of air, we can find the virtual temperature (using 1.12.4), and then using the modified equations (1.12.1)-(1.12.3) we find the virtual temperature and the pressure at the sea level.

We assume that at a given altitude the temperature, the density and the pressure have standard values that correspond to that altitude.

For example at 1000 meters, the standard values of the virtual temperature, density and pressure, obtained using (1.12.1)-(1.12.3) are respectively: 282.75 Kelvin (9.66 degree Celsius); 1.093kg/m; 665.50mm Hg.

Ballistics Coefficient in High Mountain

The form coefficient (defined by formula (2.1.5), ref. "Exterior Ballistics with Applications),

$$C(v/a) = iC_D(v/a),\qquad\qquad (1.12.5)$$

as a function of the Mach number does not change with the change of the characteristics of the atmosphere since it does not depend on those characteristics (if the atmosphere characteristics at the sea level are standard). For a non-standard atmosphere, the drag coefficient changes and depends on the speed of sound and the projectile speed (see section 8.4, "Exterior Ballistics with Applications"). Though the form coefficient is considered a constant, the definition (1.12.5) shows that it depends on the Mach number.

When shooting is in a high mountain, there is a small change in the BC of projectile since the interval of the projectile speed changes with respect to the sea level shooting (because for the same departure angle the horizontal range of impact will be somewhat greater than the respective horizontal range at the sea level).

On the other hand, for a given horizontal range on high mountain shooting, the departure angle is smaller than the respective departure angle for shooting at the same horizontal range at the sea level, and normally we would expect a slight change in "the fixed value" of the form factor, and so in the value of the ballistics coefficient.

Thus, for the shooting in high mountains the changes in the range of fire or in the angle of fire might be reflected as small change in the "predetermined" value of the sea level BC.

Anyhow, based on the information given in section 1.3, we still can use the fixed BC, predetermined experimentally at the sea level, in a standard atmosphere.

Appropriate Ballistics Coefficient

The best way to study the projectile trajectory is the use of PC programs that have the form factor or the ballistics coefficient expressed as functions of projectile speed, or as a function of the departure angle.

To find an appropriate BC, even when the firing tests are performed in high mountains, we can use the PC program Coeff.bas, or AngleC. bas on condition that the atmospheric characteristics at the altitude of shooting are standard, i.e. those characteristics correspond to the values computed using formulas (1.12.1)-(1.12.3), or close to them.

If there are deviations from the standard values at the sea level (and so in other altitudes) than the data of shooting should be converted to the "standard atmosphere" using the method presented in chapter 6 of "Exterior Ballistics with Applications".

Another way to estimate the BC based on the firing test is to guess the BC using the PC program AngleC.Bas and the data of experimental shooting, trying to match possibly all the elements of the trajectory at a given range, especially the time of flight and the terminal speed.

To make easy the calculation of BC using AngleC.Bas we can estimate in advance the BC using the PC program Coeff.Bas.

In table 2 are displayed the firing data for the 0.30 Ball M2 bullet, when the shooting take place in high mountains respectively on 500m, 1000m, and 1,500m.

The data are calculated using the PC program AngleM30.Bas. Table 2 can also be obtained using the PC program AngleC.Bas and the ballistics coefficient BC = 3.248.

Table 2—Caliber 0.30 Ball M2, Initial Speed 853.44m/s. Standard Atmosphere

Range m	Departure Angle (degree)	Cross wind 5m/s. Deflect.	Departure Angle (degree)	Cross wind 5m/s. Deflect.	Departure Angle (degree)	Cross wind 5m/s. Deflect.
	500m Altitude		1000m Altitude		1500m Altitude	
100	0.0409	0.027	0.0408	0.025	0.0407	0.024
200	0.0872	0.112	0.0867	0.106	0.0862	0.101
300	0.1400	0.265	0.1385	0.251	0.1372	0.237
400	0.2002	0.497	0.1974	0.468	0.1948	0.441
500	0.2697	0.817	0.2648	0.769	0.2603	0.723
600	0.3505	1.240	0.3424	1.164	0.3349	1.091

Example 1.32

The BC of the Caliber 0.30 Ball M2 bullet fired with initial speed 853.44m/s determined at the sea level in standard atmosphere, for the distance 500 meters is 3.248 (see table 1, section 1.11). Departure angle is 0.275 degree. The time of flight is 0.76 seconds, the terminal speed is 509.20m/s, and the vertex is located at the point with coordinates (272m, 0.71m).

To verify the validity of the BC of the same bullet the firing tests were performed in a mountain at the altitude 1500 meters, for the same horizontal range. The departure angle needed to hit the center of the target at 500 meters was 0.261 degree.

Use the PC Coeff.Bas and AngleC.Bas to verify that the same BC is valid for the shooting at altitude 1500 meters.

Solution

Using the PC Coeff.Bas

Input: Guessed BC, 3.2; Initial Speed, 853.44; Departure angle, 0.26055; Range, 500; y-coordinate of Target, 1500, y-coordinate of Gun, 1500; Error, 0.8; Step, 1;
Output: BC, 3.248, speed 549.50m/s; time 0.73s.

Remark

At the altitude 1500 meters over the sea level, firing the bullet with departure angle 0.275 degree we obtain a range of 519 meters, while at the horizontal distance 500 meters from the gun the bullet will pass around 13 centimeters over the center of the target.

1.13 Maximum Range and Vertical Shooting

For safety purposes during shooting with military rifles or hunting arms it is necessary to know the maximum range, the kinetic energy of the bullet at the maximum range and, in general, the elements of projectile trajectory in large distances from the firing point.

As far as I know, in the ballistics literature, there are no satisfying theoretical methods to estimate with acceptable accuracy the maximum range of a bullet fired from a rifle.

The military bullets are usually tested, and the results are present in different army manuals.

Robert L. McCoy, in his wonderful book "Modern Exterior Ballistics", Schiffer Publishing Ltd, 1999, has estimated the maximum ranges of some projectiles, but the way they are estimated is not shown though we might assume that they are result of theoretical analyses and practical tests.

We will use the results of McCoy for the bullet caliber 0.30 Ball M2 to find a model to estimate the elements of the projectile trajectory.

The results we obtain are only theoretical estimations obtained for the standard atmosphere with no assurance for the degree of accuracy. In other words, the results we have obtained need to be verified in practice of shooting.

Maximum Range

In **Exercise 1.8** (Chapter 1) we found that the Siacci Ballistics Coefficient of caliber 0.30 Ball M2 bullet (fired with speed 853.44m/s) that corresponds to the maximum range is $c = 4.8439$, while the average ballistics coefficient is $c = 4.93176$.

Using the PC program RangeC.Bas, we find the following results for the elements of the trajectory of the bullet fired with an angle 32 degree.

- For the value of the ballistics coefficient $c = 4.8439$ the elements of the trajectory at the impact point are: Range, 3141m; time of flight, 25.71s; speed, 110m/s; terminal angle, -65.04704 degree; trajectory vertex, (2007m, 886m).
- For the value of the ballistics coefficient $c = 4.93176$ the elements of the trajectory at the impact point are: Range, 3100m; time of flight, 25.71s; speed, 110m/s; terminal angle, -65.04704 degree; trajectory vertex (1983, 877).

Remark

Using the average coefficient we get an approximate value for the maximum Range (3100m) that is 41 meters less than the actual range (3141m).

Maximum Altitude for Vertical Shooting

In the following calculations, we use the average ballistics coefficient $c = 4.93176$. Employing the PC program RangeC.Bas for a value of departure angle of 89.6 degree (i.e. for an approximate vertical shooting), we get the following values for the elements of the trajectory:

Maximum altitude, 2310m; time of flight, 43.70s; impact speed, 126.80m/s.

If the departure speed of the 0.30 Ball M2 bullet is 823m/s (2700fps) then (using the PC program RangeC.Bas with $c = 4.93176$), for the elements of the trajectory of the bullet fired approximately vertically (departure angle 89.6 degree) we obtain the following elements of the trajectory at the impact point:

Maximum altitude, 2280m; time of flight, 43.50s; impact speed, 126.70m/s

Comparing the impact speed of the bullet fired vertically with initial speed 823m/s with the impact speed of the bullet fired vertically with speed 853m/s, we see that they are almost identical.

It means that the impact speed of the same bullet is the terminal speed determined by the ballistics coefficient (bullet form factor). This conclusion is well matched with the Schaefer's statement: "If I remember correctly from my limited parachuting experience the terminal velocity of a falling person is somewhere around 130 mph, or about 200 f/s.", ref. *http://www.frfrogspad.com/miscella.htm#straight*.

Remark

To find the trajectory vertex we need to do the following modification in the RangeC.Bas:

Substitute the instruction:

IF ABS(z0) < .0005 THEN 'Condition for maximum height, tan = 0,

with the instruction:

IF ABS(z0) < .5 THEN 'Condition for maximum height, tan = 0

Kinetic Energy

The kinetic energy of the bullet at the maximum range (departure angle 32 degree is

$$K = \frac{mv^2}{2} = \frac{(0.0097) \square (110)^2}{2} = 58.70 \, Joule$$

The kinetic energy at the point of impact of the bullet fired vertically is

$$K = \frac{mv^2}{2} = \frac{(0.0097) \square (127)^2}{2} = 78 \, Joule$$

Remark

- The kinetic energy of a bullet falling vertically on ground (78 joule) is equal to the threshold lethal energy a projectile, while the kinetic energy of the bullet at the maximum range (3400m) is somewhat less then the threshold of lethal kinetic energy.

The estimated values of kinetic energy shows that the bullet caliber 0.30 M2 falling vertically, or reaching the maximum range, still has enough energy to lethally injure the person it strikes.

These conclusions are compatible with Hatcher's experiments for the same bullet (see *http://www.frfrogspad.com/miscella.htm#straight*).

- The estimation of the impact speed and kinetic energy is based on the assumption that the bullet fired vertically maintains its gyroscopic stability during the flight to the ground.

In a private communication, John Schaefer (http://www.frfrogspad.com) writes:

"The range of KE needed to "injure" runs from about 50 ft/lbs to 150 ft/lb (68-203 Joule). The range is so great because of the properties of skin and where the projectile strikes determines the break point. Interestingly projectile shape has very little effect.

Vertically fired bullets tend to come down base first due to the gyroscopic stability (Typical .30 cal. bullet is rotating in excess of 175,000 rpm at the muzzle and spin decays very slowly). The final velocity of the bullet is limited by air drag".

1.14 QBasic PC Program COEFF.BAS

The PC program Coeff.bas can be used to find the ballistics coefficient of a projectile when the muzzle of the firearm is at the sea level and we know the horizontal range (or the coordinates of the terminal point), the departure angle, and the departure speed. The projectile flight is in standard atmosphere.

The ballistics coefficient estimated using Coeff.Bas depends on the "error" we input and might slightly vary as a function of error.

The results obtained employing Coeff.Bas include also the projectile speed at the terminal point.

Using Coeff.Bas

Example 1.33 Standard Atmosphere, Sea Level

A bullet caliber 0.224, 5.56mm M855 (62 grain) fired horizontally (departure angle 0 degree) with speed 940m/s hits the target (firing board), located at the horizontal range 500 meters, 2.197 meters below the horizontal range. Terminal speed is 483m/s; time of flight, 0.743s.

Find the Siacci ballistics coefficient of the given bullet.

Solution

First, we find the departure angle:

$$\alpha_0 = \frac{drop}{x} = \frac{2.197}{500} = 0.004394 = 0.004394 \cdot (\frac{180}{\pi}) = 0.251757°$$

Execute the PC program Coeff.Bas.
Would you like to cancel the data file? Press "Enter" key.
Guess the initial coefficient, Input: 4.1 "Enter".

Input: Projectile Speed, 940; departure angle, 0.251758; x-coordinate of target, 500; Y-coordinate of Target, 0; y-coordinate of the gun, 0; Error in x-coordinate, 0.5; Number of Steps, 1;

Output: Ballistics coefficient, **4.247**; time, 0.743; terminal speed, 484.6; error in y-coordinate 0.0034m.

Note: If we input another error in x, we find another value of BC. For example, if the error in x is 0.8m, then we find a BC of 4.239; if the error is 0.02

PC program Coeff.Bas
Standard Atmosphere
Finding Ballistics Coefficient Automatically

'Ballistics Coefficient : c=1000id^2/m
'Standard Atmosphere
'The computed coefficient "ko" is reserved in the file c:/koef.dat
'_____

'Control Example
'Input Data
'Cancel data File : Print y/n

'INPUT "Guess Initial Coefficient =   "; 0.2
'INPUT "Departure Angle [Degree.Minutes] "; 6.2
'INPUT "Initial speed [m/s]   "; 885
'INPUT "x-coordinate of Target [m] =   "; 10000
'INPUT "Y-coordinate of Target [m]   "; 0
'INPUT "y-coordinate of GUN   "; 0
'INPUT "Number of Steps n   "; n "Input: 100
'Input "Error in x-coordinate "; gab "Input: 1

'Output:
'Ballistics Coefficient BC   = 0.2389
'X-coordinate of Target/Range = 10000
'Departure angle   = 6.2
'Impact angle   = - 10.19137
'Error in y-coordinate   = 0.1798
'_____

'Functions, Subs

DECLARE SUB y1z1v1w1 (x, y, z, v, w, y1, z1, v1, w1, koef)
DECLARE SUB InfHyres (koef, kk, dis, n, speed, gab, yy, vo)
DECLARE SUB NPxyzvw (nk, x, x0, y, y0, z, z0, v, v0, w, w0, h, h0, k, l, r, q)
DECLARE SUB NPkoef (k, l, r, q, h, y1, z1, v1, w1)
DECLARE SUB menu (cog, cof, xf, yf, xfu, yfu, t$)
DECLARE SUB Rezervim (koef, kek, gab, x0)
DECLARE SUB KthimiKendit (kk)

'Variables

```
DIM m(4, 4), v(4)
rendi = 4
cog = 7: cof = 0
gab = gab
dkoef = .0001
menu cog, cof, 3, 10, 21, 70, "INPUT"
InfHyres koef, kk, dis, n, speed, gab, yy, vo
CLS
PRINT "First Angle="; kk
hap = n
KthimiKendit kk

'Solution
ff:
x0 = 0: y0 = kk: z0 = y0: v0 = vo: w0 = 0: xx = dis: h0 = hap
y0 = speed * COS(y0 * 3.141516954# / 180)
z0 = TAN(z0 * 3.141516954# / 180)

f:
FOR nk = 1 TO rendi
NPxyzvw nk, x, x0, y, y0, z, z0, v, v0, w, w0, h, h0, k, l, r, q
y1z1v1w1 x, y, z, v, w, y1, z1, v1, w1, koef
NPkoef k, l, r, q, h, y1, z1, v1, w1
m(nk, 1) = k: m(nk, 2) = 1
m(nk, 3) = r: m(nk, 4) = q
NEXT nk

FOR i = 1 TO rendi
v(i) = 1 / 6 * (m(1, i) + 2 * m(2, i) + 2 * m(3, i) + m(4, i))
NEXT i

'New Point
x0 = x0 + h: y0 = y0 + v(1): z0 = z0 + v(2)
v0 = v0 + v(3): w0 = w0 + v(4)

IF ABS(x0 - xx) <= .01 AND v0 <= (yy + gab * TAN(kk * 3.1415 / 180)) AND
    v0 >= (yy + (-1 * gab) * TAN(kk * 3.1415 / 180)) AND ABS(x0 - xx) <= .01
    THEN
'Display results
```

```
kek = gab * z0
impact = 180 * ATN(z0) / 3.141592654#
CLS
LOCATE 7, 21: PRINT "Ballistics Coefficient BC = "; koef
LOCATE 8, 21: PRINT "Range        = "; x0
LOCATE 9, 21: PRINT "Departure Angle = "; kk
LOCATE 10, 21: PRINT "Impact Speeed = "; y0
LOCATE 11, 21: PRINT "Time of Flight = "; w0
LOCATE 12, 21: PRINT "Impact Angle  = "; impact
LOCATE 12, 21: PRINT "Error in y-coordinate = "; ABS(kek)
LOCATE 13, 21: PRINT "Error in x-coordinate = "; ABS(gab)
PLAY "a8a16b8a8"
Rezervim koef, kek, gab, x0
dis = dis + 200
IF dis > disf THEN PRINT "END": INPUT b: GOTO fundi:
hap = n
READ kk
PRINT "New Angle ="; kk
KthimiKendit kk
GOTO ff:
fundi:
END
END IF


dkoef = .0001
IF ABS(x0 - xx) <= .01 AND v0 > (gab * TAN(kk * 3.1415 / 180)) THEN
koef = koef + dkoef
PRINT v0, koef
GOTO ff:
END IF
IF ABS(x0 - xx) <= .01 AND v0 < ((-1 * gab) * TAN(kk * 3.1415 / 180)) THEN
koef = koef - dkoef
PRINT v0, koef
GOTO ff:
END IF
GOTO f:
END


SUB InfHyres (koef, kk, dis, n, speed, gab, yy, vo)
```

```
CLS
LOCATE 5, 12: INPUT "Would You Like to Cancel the data File [Y/N]"; y$
IF y$ = "y" OR y$ = "Y" THEN KILL "c:\koef.dat"
CLS
LOCATE 5, 12: INPUT "Guess Initial Coefficient  = "; koef
LOCATE 7, 12: INPUT "Projectile Speed       = "; speed
LOCATE 8, 12: INPUT "Launching Angle        = "; kk
LOCATE 10, 12: INPUT "x- coordinate of target  = "; dis
LOCATE 11, 12: INPUT "Y - coordinate of Target  = "; yy
LOCATE 12, 12: INPUT "y - coordinate of the gun = "; vo
LOCATE 14, 12: INPUT "Error in x-coordinate    = "; gab
LOCATE 16, 12: PRINT "Number of Steps when range Ends with  0   is [10]"
LOCATE 17, 12: PRINT "Number of Steps when range Ends with 00  is [100]"
LOCATE 18, 12: PRINT "Number of Steps range Ends with a NON ZERO
    Number is [1]"
LOCATE 20, 12: INPUT "Number of Steps        = "; n
END SUB


SUB KthimiKendit (kk)
kk = kk
END SUB


SUB menu (cog, cof, xf, yf, xfu, yfu, t$)
COLOR cog, cof
LOCATE xf - 1, yf: PRINT t$
LOCATE xf, yf: PRINT "É" + STRING$(yfu - yf, 205) + "»";
FOR i = xf + 1 TO xfu
LOCATE i, yf: PRINT "º" + SPACE$(yfu - yf) + "º";
NEXT
LOCATE xfu + 1, yf: PRINT "È" + STRING$(yfu - yf, 205) + "¼";
END SUB


SUB NPkoef (k, l, r, q, h, y1, z1, v1, w1)
k = h * y1: l = h * z1
r = h * v1: q = h * w1
END SUB


SUB NPxyzvw (nk, x, x0, y, y0, z, z0, v, v0, w, w0, h, h0, k, l, r, q)
IF nk = 1 THEN
```

```
x = x0: y = y0: z = z0
v = v0: w = w0: h = h0
GOTO fund:
END IF
IF nk = 2 OR nk = 3 THEN
x = x0 + (.5 * h): y = y0 + (.5 * k)
z = z0 + (.5 * l): v = v0 + (.5 * r)
w = w0 + (.5 * q)
GOTO fund:
END IF
IF nk = 4 THEN
x = x0 + h: y = y0 + k: z = z0 + l
v = v0 + r: w = w0 + q
END IF
fund:
END SUB


SUB Rezervim (koef, kek, gab, x0)
OPEN "c:\koef.dat" FOR APPEND AS #1
PRINT #1, koef, kek, gab, x0
CLOSE #1
END SUB


SUB y1z1v1w1 (x, y, z, v, w, y1, z1, v1, w1, koef)
IF (y * SQR(1 + z ^ 2)) >= 256! THEN
y1 = -1 * koef * ((289.08 - .006328 * v) / 289.08) ^ 4.4 * (1 / 3 - 80 / (y *
    SQR((1 + z ^ 2))))
ELSE
y1 = -1 * koef * ((289.08 - .006328 * v) / 289.08) ^ 4.4 * .0001212 * y *
    SQR((1 + z ^ 2))
END IF
z1 = -9.80665 / y ^ 2
v1 = z
w1 = 1 / y
END SUB
```

2

COMPUTATION OF THE DEPARTURE ANGLE

Introduction

The computation of the departure angle of a projectile needed to hit a given target is the main problem we face in the practice of shooting. The PC programs included in this chapter can be used to find the departure angle when a projectile is launched with a given initial speed and the range of fire or the coordinates of the gun and the target are known. The ballistics coefficient is known as well.

For all PC programs, we assume that the origin of the coordinates is at the sea level.

The PC programs presented hereafter can be classified into three categories:

- PC programs that use constant BC for any trajectory of the given projectile;
- PC programs that employ a BC that is function of the projectile speed;
- PC programs that use a BC that is a function of the departure angle.

The PC programs can be grouped in two other categories, based on the fact that they are used when the flight is in standard atmosphere, or a non-standard atmosphere, and if the projectile has standard characteristics, or not.

Using the PC programs presented in this chapter, together with the PC program Coeff.Bas (chapter 1) and the programs presented in chapter 2, the reader is able to solve a large variety of exterior ballistics problems encountered in the theory and practice of shooting with small arms or other non-reactive firearms.

In each section through examples and exercises, we illustrate the use of the respective PC program to solve different exterior ballistics problems.

The PC programs are shown at the end of each section.

2.1 PC Program Anglec.Bas, Inclined Shooting

The PC program **Anglec.Bas** can be used to determine the departure angle of a projectile needed to hit a given target in the Standard Atmosphere (in presence or not of the wind), when are known: "The ballistics coefficient of the projectile fired, the projectile speed, the coordinates of the target (impact point) and the location of the firearm". The program estimates the coordinates of the trajectory vertex, as well as the time of flight to the target, the terminal speed and the terminal angle, the crosswind deflection of the projectile.

The PC program Anglec.Bas finds as well the trajectory elements of a point on the trajectory for a given x-coordinate of that point. The value of x-coordinate of the point must be less than the projectile range.

The **Anglec.Bas** program uses a constant Siacci's ballistics coefficient that represents an average coefficient that usually is given in ballistics tables, or in the firearm-bullet handbook (See the information on BC given in Chapter 1).

The PC program **Anglec.Bas** (see below) can be used for anti-aircraft or uphill shooting, as well as for downhill shooting, in standard atmosphere.

For uphill shooting, the program output for the trajectory vertex is (0, 0). That means that the trajectory vertex "does not exists" and the largest altitude of the trajectory is at the point of impact.

Mathematically, the PC program is unable to find the vertex that is programmed to be found as the point on the trajectory where the tangent is parallel to the x-axis.

Use of AngleC.Bas

Example 2.1

The departure speed of a122mm cannon is 885m/s. The ballistics coefficient of the projectile is 0.2389.

(a) Find the departure angle needed to hit a target located at the sea level at the horizontal range 10,000m. The cannon are at the sea level. The atmosphere is standard and no wind is present.
(b) Find the departure angle needed to hit the target located at the sea level at the horizontal range 10,000m. The atmosphere is standard but there is a range wind of 10m/s.
(c) Find the departure angle needed to hit a target located at a range of 10,000 meter and 200 meters over the sea level, i.e. the coordinates of the target are (10000m, 200m). The coordinates of the cannon are (0m, 60m). Find as well the elements of the trajectory at a point with abscissa 5500m. The atmosphere is standard and no wind is present.

Solution

(a) **Input**: x-coordinate of target, 10000; y-coordinate of target, 0; y-coordinate of cannon, 0; departure speed, 885m/s; BC, 0.2389.
   Output: The program gives the following results: Departure angle 6.20 Degree, time of flight 16.75, terminal angle = -10.19112 degree, coordinates of the trajectory vertex respectively x=5631m and y= 347m.
(b) In the same way as in (a), but inputting the range wind 10m/s, we find that the departure angle is 6.12 degree.
(c) **Input**: x-coordinate of target, 10000; y-coordinate of target 200, y-coordinate of cannon, 60; departure speed, 885m/s; BC, 0.2389.

Input as well the abscissa of a point on the trajectory, for example 5500m.

Output: The program gives the following results:

Departure angle 6.976685 Degree, time of flight 16.70, terminal angle = -09.3203 degree, coordinates of the trajectory vertex respectively x=6100m and y= 487m.

For the trajectory point with abscissa x =5500m we obtain the following values:

The ordinate, 482 meter; the time of flight, 7.62 seconds; the projectile speed, 594.50m/s; the angle of flight, 0.9991 degree.

Exercise 2.2

Use the BC = **4.132** found in Exercise 1.5, chapter 1, to find the departure angles of the 7.62mm Russian bullet, Mod. 1943, mass 0.0079kg respectively for ranges 300m and 800m, as well as all the other elements of the trajectory at the impact point. Projectile speed is 735m/s.

Projectile is fired in standard atmosphere, the muzzle of the firearm and the target are at the sea level. The speed of the cross wind is 5m/s.

Answer

Range 300m: Angle is 0.206421 degree; Time is 0.50s; Speed is 487.80m/s;

Impact angle is -0.27276; Vertex (160m, 0.31m). Cross-wind Deflection, 0.471m.

Range 800m: Angle is 0.983887 degree; Time is 1.94s; Speed is 274.40m/s;

Impact angle is -1.88022; Vertex (467m, 4.86m); Cross-Wind Deflection, 4.255m.

Note: For comparison hereafter are given the same data from the Range Table of the 7.62mm Russian projectile.

Range 300m: Angle is 0.216 degree; Time is 0.50s; Speed is 485.00m/s;

Impact angle is -0.276; Vertex (161m, 0.31m).

Range 800m: Angle is 1.020 degree; Time is 1.96s; Speed is 276.00m/s;
Impact angle is -1.980m/s; Vertex (462m, 5.10m).

Exercise 2.3

In Example 5 (Ref. G. Klimi, "Exterior Ballistics with Applications", p. 97-98, Xlibris 2008) for the bullet caliber 0.30 Ball M2 we found that the average BC of 3.385.
Use BC = 3.385 to find the departure angle needed to hit a target located at a range 549 meter (600 yard) if the speed of the bullet is 853.44m/s.
Answer: 0.3208 Degree = 19.25 Minute

Example 2.4

Find the departure angle needed to hit a target located at the point with coordinates (500m, 866m) if the departure speed of the projectile is 945m/s. The BC is 1.64. The projectile trajectory is in standard atmosphere and the departure point is at the sea level.

Solution

Input: The abscissa of the target 500, the ordinate of the target 866, the departure speed 945, the BC is 1.64.
The results obtained executing the programs are:
The departure angle is 60.22168 degree; time of flight, 1.355 seconds; projectile terminal speed, 579.06m/s; terminal angle, 59.69694 degree. Distance to the target, 1000m.

Exercise 2.5 Rifleman's Rule

The departure speed of a 5.56mm M855 bullet is 940m/s. The Siacci ballistics coefficient of the bullet is BC is 4.246 (see section 1.1).

(a) Find the departure angle needed to hit a target located at the point with coordinates (300m, 250m). The muzzle of the rifle is located at the point with coordinates (0m, 125m), i.e. 125 meter above the sea level. Find as well the super elevation angle.

(b) Find the departure angle if the rifle is at the sea level, while the target is 125 meters over the sea level at the point with coordinates (300, 125).

(c) Find the departure angle needed to zero the rifle at 300 meters if the target and the rifle are at the sea level.

Answer: (a) Departure angle is 22.74219 degree (The site angle is 22.619895 and the super elevation angle is 0.12233 degree, or 7.3395MOA), distance to the target 325 meters. (b) The departure angle is 22.74219 degree, (The site angle is 22.619895 degree, and the super elevation angle is 0.12233 degree or 7.3395 MOA. (c) The departure angle is 0.123169 degree or 7.39 MOA.

Rifleman's Rule

Comparing the results in (a), (b) and (c) we see that the super elevation angle in (a) and (b) and the departure angle in (c) are practically the same:

$$\bar{\alpha}_0 \approx \alpha_0 = 7.39 MOA .$$

At the same time, we see that the horizontal range is the product of the inclined range and the cosine of site angle:

$$x = D \cdot \cos(E) = 325 \cdot \cos(22.619895) = 300m .$$

The results are compatible with Rifleman's rule studied in section (1.6).

Exercise 2.6

The Siacci ballistics coefficient of 7.62mm M80 bullet (see section 1.1) is 3.182. The bullet is fired with a speed of 856.50m/s.

(a) Find the departure angle needed to zero the gun at the horizontal range 300 meters.
(b) Find the departure angle needed to zero the rifle at 300 meters if the powder load is reduced and the bullet speed is 800m/s.
(c) Find the departure angle needed to hit a target located at the point with coordinates (500m, 600m), if the bullet is fired at a speed 856.50m/s. The bullet departure point is 750 meter over the sea level.

Answer: (a) 0.13928 degree (8.36MOA); (b) 0.16144 degree (9.69MOA); (c) Distance to target 522m, departure angle -16.46298 degree (Super elevation angle is -0.23626 degree (or -14.175 MOA).

QBasic PC Program ANGLEC.BAS
Standard Atmosphere, Wind Present

```
'FIND: LAUNCHING Angle, Time of Flight, y-coordinate of a point, etc.
'GIVEN: Coordinates of target and the gun, Ballistics Coefficient, Launching
    Speed
        '(Standard Atmosphere), Range-Wind, Cross-Wind are Present
'_____

'CONTROL DATA
'INPUT: Range = 10,000: Launching Speed = 885: Ballistics Coefficient
    c = 0.2389,
'x-coordinate of a point on Trajectory = 5500
'

'RESULTS: Launching Angle = 6.20 [Degree], Time of Flight = 16.75 [Sec.]
'Terminal Angle = -10.191 [Degree]
'Coordinates of Trajectory vertex (5631, 347)
'For x =5500m: y = 346.63m, Time = 7.62s,
'Angle = 0.2119 Degree, Speed = 593.20m/s
'_____

'Note: Round the Input RANGE to the nearest 1; neglect the numbers after
    decimal point
'_____

                'Functions and Sub. Prog.


DECLARE SUB y1z1v1w1 (x, y, z, v, w, y1, z1, v1, w1, koef, ys, yy, wind)
DECLARE SUB InfHyres (xx, n, koef, vv, vo, yo, xc1, cw, wind)
DECLARE SUB InfDales (x0, y0, z0, v0, w0, xm, ym, A, xc, yc, tc, vc, ac,
    vo, vv, xx, vo, cw)
DECLARE SUB NPxyzvw (nk, x, x0, y, y0, z, z0, v, v0, w, w0, h, h0, k, l, r, q)
DECLARE SUB NPkoef (k, l, r, q, h, y1, z1, v1, w1)
DECLARE SUB menu (cog, cof, xf, yf, xfu, yfu, t$)
DECLARE SUB y0z0 (y0, z0, A, vo, wind)
DECLARE SUB c (koef, y0, A)

'Variables
SCREEN 0
1 :
DIM m(4, 4), v(4)          'Intermediate values (k,l,r,q)
rendi = 4                  'rend dif.
cog = 7: cof = 0
cikli = 0
```

```
A = 23                          'Initial Angle 23 degree
kendi = 22                      'd.Angle for maximum distance
kov = 1                         'Test of the value of v0
tt = 1

'Solution
CLS
'Initial Data
menu cog, cof, 3, 10, 7, 70, "DATA INPUT"
InfHyres xx, n, koef, vv, vo, yo, xc1, cw, wind
hap = 1

IF wind = 0 THEN
gab = .01
ELSE
gab = 1
END IF

'Initial values
f:
x0 = 0: v0 = vv: w0 = 0
y0z0 y0, z0, A, vo, wind: h0 = hap
c koef, y0, A
ff:
FOR nk = 1 TO rendi
NPxyzvw nk, x, x0, y, y0, z, z0, v, v0, w, w0, h, h0, k, l, r, q
y1z1v1w1 x, y, z, v, w, y1, z1, v1, w1, koef, ys, yy, wind
NPkoef k, l, r, q, h, y1, z1, v1, w1
m(nk, 1) = k: m(nk, 2) = l
m(nk, 3) = r: m(nk, 4) = q
NEXT nk

'Estimations for new points
FOR i = 1 TO rendi
v(i) = 1 / 6 * (m(1, i) + 2 * m(2, i) + 2 * m(3, i) + m(4, i))
NEXT i

'New Points
x0 = x0 + h: y0 = y0 + v(1): z0 = z0 + v(2)
v0 = v0 + v(3): w0 = w0 + v(4)
```

```
xcc = x0 + wind * w0
IF ABS(xcc - xc1) <= .5 THEN
xc = xcc

yc = v0
tc = w0
ac = (180 / 3.141592654#) * ATN(z0)
vc = y0 / COS(ATN(z0))
END IF

IF xmm > 30 AND ABS(z0) <= .00001 THEN
xm = xmm

ym = v0
END IF

'Tests the y-value
xT = x0 + wind * w0
IF kov = 1 THEN kov = -1: GOTO ff:
IF ABS(xT - xx) < gab AND v0 <= yo + (gab * TAN(A * 3.1415954# / 180))
    AND v0 >= yo + ((-1 * gab) * TAN(A * 3.1415954# / 180)) THEN
c:
'Display Results
CLS
PLAY "a8a16a32b8"
menu 12, 0, 5, 10, 11, 70, "RESULTS:"
COLOR 7
InfDales x0, y0, z0, v0, w0, xm, ym, A, xc, yc, tc, vc, ac, yo, vv, xx, vo, cw
CLS
GOTO 1:
END IF

IF ABS(xT - xx) < gab AND v0 > yo + (gab * TAN(A * 3.1415954# / 180)) THEN
t$ = "* ? *"
menu 18, 0, 10, 20, 14, 60, t$
COLOR 14
LOCATE 12, 30: PRINT "Wait a moment, Please (+)";
LOCATE 12, 53: PRINT tt
tt = tt + 1
```

```
COLOR 7
A = A - kendi
GOTO fff:
END IF

IF ABS(xT - xx) < gab AND v0 < yo + ((-1 * gab) * TAN(A * 3.1415954# / 180))
    THEN
t$ = "* ? *"
menu 18, 0, 10, 20, 14, 60, t$
COLOR 14
LOCATE 12, 30: PRINT "Wait a moment, Please (-)";
LOCATE 12, 53: PRINT tt
tt = tt + 1
COLOR 7
A = A + kendi
GOTO fff:
END IF
GOTO ff:

fff:
'Restart Cycle
cikli = cikli + 1
IF cikli = 20 THEN GOTO c:
kendi = kendi / 2
kov = 1
GOTO f:

SUB c (koef, y0, A)
koef = koef
END SUB

SUB InfDales (x0, y0, z0, v0, w0, xm, ym, A, xc, yc, tc, vc, ac, yo, vv, xx, vo, cw)
aT = ATN(z0) * 180 / 3.141592654#
LOCATE 6, 16: PRINT "Launching Angle    = "; A; "degree"
LOCATE 7, 16: PRINT "Time of Flight   = "; INT((w0) * 1000 + .5) / 1000; "seconds";
LOCATE 8, 16: PRINT "Terminal Speed     = "; INT((y0 / COS(ATN(z0)))
    * 100 + .5) / 100; "m/s"
LOCATE 9, 16: PRINT "Terminal Angle    = "; aT; "degree"
LOCATE 10, 16: PRINT "Trajectory Vertex = :"; "("; INT((xm) * 100 + .5) /
    100; ","; INT((ym) * 100 + .5) / 100; ")"
```

```
LOCATE 11, 16: PRINT "Cross-Wind Deflection    = "; INT((cw * (w0
    - xx / (vo * COS(A * 3.14159265# / 180)))) * 1000 + .5) / 1000
LOCATE 13, 16: PRINT "Distance to the target    = "; INT(((xx ^ 2 + (vv
    - v0) ^ 2) ^ .5) * 100 + .05) / 100; "meter";
LOCATE 14, 16: PRINT "x, y coordinates of TARGET = "; "("; INT((xx) *
    100 + .5) / 100; ","; INT((v0) * 100 + .5) / 100; ")"
LOCATE 15, 16: PRINT "x,y coordinatees of GUN    = "; "("; 0; ","; vv; ")"
LOCATE 17, 18: PRINT "x-coordinate of a point[m] :"; INT((xc) * 100 + .5) / 100
LOCATE 18, 18: PRINT "Corresponding y [m]      :"; INT((yc) * 1000 + .5) / 1000
LOCATE 19, 18: PRINT "Corresponding Time [sec]  :"; INT((tc) * 100 + .5) / 100
LOCATE 20, 18: PRINT "Corresponding Speed [m/s] :"; INT((vc) * 100 + .5) / 100
LOCATE 21, 18: PRINT "Corresponding Angle [Deg]  :"; ac
LOCATE 22, 18: PRINT "Cross-Wind Deflection  :"; INT((cw * (tc - xc / (vo * COS(A
    * INT((cw * (tc - xc / (vo * COS(A * 3.14159265# / 180)))) * 1000 + .5) / 1000
COLOR 7
LOCATE 24, 11: PRINT "Pres [ P ] to repeat [ Esc ] to end";
cc$ = INPUT$(1)
IF cc$ = CHR$(27) THEN SCREEN 9: CLS : END
END SUB


SUB InfHyres (xx, n, koef, vv, vo, yo, xc1, cw, wind)
LOCATE 4, 13: INPUT "x-coordinate of TARGET [m]   ="; xx
LOCATE 5, 13: INPUT "y-coordinate of TARGET [m]   ="; yo
LOCATE 6, 13: INPUT "y-coord of GUN               ="; vv
LOCATE 7, 13: INPUT "Initial Speed [m/s]        ="; vo
LOCATE 9, 13: INPUT "X-coordinate of a point     ="; xc1
LOCATE 10, 13: INPUT "Ballistics Coefficient    = "; koef
LOCATE 11, 13: INPUT "Range-Wind                = "; wind
LOCATE 12, 13: INPUT "Cross-Wind                = "; cw
CLS


END SUB
SUB menu (cog, cof, xf, yf, xfu, yfu, t$)
COLOR cog, cof
LOCATE xf - 1, yf: PRINT t$
LOCATE xf, yf: PRINT "É" + STRING$(yfu - yf, 205) + "»";
FOR i = xf + 1 TO xfu
LOCATE i, yf: PRINT "º" + SPACE$(yfu - yf) + "º";
NEXT
LOCATE xfu + 1, yf: PRINT "È" + STRING$(yfu - yf, 205) + "¼";
```

```
END SUB

SUB NPkoef (k, l, r, q, h, y1, z1, v1, w1)
k = h * y1: l = h * z1
r = h * v1: q = h * w1
END SUB

SUB NPxyzvw (nk, x, x0, y, y0, z, z0, v, v0, w, w0, h, h0, k, l, r, q)
IF nk = 1 THEN
x = x0: y = y0: z = z0
v = v0: w = w0: h = h0
GOTO fund:
END IF
IF nk = 2 OR nk = 3 THEN
x = x0 + (.5 * h): y = y0 + (.5 * k)
z = z0 + (.5 * l): v = v0 + (.5 * r)
w = w0 + (.5 * q)
GOTO fund:
END IF
IF nk = 4 THEN
x = x0 + h: y = y0 + k: z = z0 + l
v = v0 + r: w = w0 + q
END IF
fund:
END SUB

SUB y0z0 (y0, z0, A, vo, wind)
y0 = SQR(vo ^ 2 + wind ^ 2 - 2 * vo * wind * COS(A * 3.141592654# / 180))
y0 = y0 * COS(A * 3.141592654# / 180)
z0 = TAN(A * 3.141592654# / 180)
z0 = z0 / (1 - wind / (vo * COS(A * 3.141592654# / 180)))
END SUB

SUB y1z1v1w1 (x, y, z, v, w, y1, z1, v1, w1, koef, ys, yy, wind)
yy = y * SQR(1 + z ^ 2)
IF yy > 256! THEN
y1 = -1 * koef * ((289.08 - .006328 * v) / 289.08) ^ 4.4 * (yy - 240) / (3 * yy)
ELSE
y1 = -1 * koef * ((289.08 - .006328 * v) / 289.08) ^ 4.4 * .0001212 * yy ^ 2 / yy
END IF
z1 = -9.80665 / y ^ 2
v1 = z
w1 = 1 / y
END SUB
```

2.2 PC Program AngleM30.Bas, Bullet 0.30 Ball M2

The PC program AngleM30.Bas can be used to determine the departure angle of a caliber 0.30 Ball M2 bullet needed to hit a given target when are known:

The coordinates of the target and the gun, the projectile speed and the BC of the projectile as a function of the projectile speed (Ref.: G. Klimi, "Exterior Ballistics with Applications", p.422, Xlibris, 2008).

It can be used for anti-aircraft or uphill shooting and for the downhill shooting as well.

AngleM30.Bas is valid only for a bullet caliber 0.30 Ball M2, mass 0.00972kg when projectile is launched in the standard atmosphere and in presence or not of the range-wind and cross wind.

The program can be modified in order to use it with any projectile when the BC is known as a function of the projectile speed.

The PC program AngleM30.Bas represents a slight modification of the PC program AngleC.Bas.

Note: The PC program Antair.Bas and Dwnh1.Bas (Ref. "Exterior Ballistics with Applications", page 587) are not presented in "the companion" since AngleC.Bas and AngleM30.Bas are a better substitute.

The following examples and exercises are solved using the PC program AngleM30.Bas.

Use of AngleM30.Bas

Example 2.7 Angle of Sight

For the 0.30 ball M2 bullet, caliber 0.308 inch, departure speed 853.44m/s, if the projectile range is 366m (400 yard) find:

Horizontal Fire, Use the PC program AngleM30.Bas

(a) Find the departure angle and all the other main elements of the trajectory. The target and the rifle are 1.7 meters over the sea level. The atmosphere is standard.

Find as well the crosswind deflection if there is a cross wind 8m/s. How does the departure angle changes if there is a range wind of 10m/s?

(b) Find the angle of sight needed to zero the gun at 366 meters from the muzzle of the gun location if the sight height is 1.5 inches (0.381m) over the bore centerline.

   The atmosphere is standard and the shooting is at the sea level (y-coordinate of the gun and of the target are equal to 1.70m).

(c) Find the angle of sight needed to zero the gun at 457 meters (500 yards) from the muzzle if the sight is 1.5 inches (0.381m) over the bore centerline.

The atmosphere is standard and the shooting is at the sea level (y-coordinate of the gun and of the target are equal to 1.70m).

Inclined Fire

(d) Find the angle of sight if the site angle is 30 degree, x-coordinate of the target is 366 meters (400 yards) and the center of the target is 1.7 meter over the inclined ground. The shooter is in standing position and the muzzle is 1.7 meters over the inclined ground.

Solution

Horizontal Fire

(a) **Input**: x-coordinate of target, 366; y-coordinate of target, 1.7; y-coordinate of the gun, 1.7; projectile initial speed, 853.44, crosswind, 8m/s.

The results, obtained executing the PC program, are:
Departure angle, 0.18091 degree; time of flight, 0.516 seconds; terminal speed,590m/s, terminal angle, -0.232 degree, trajectory vertex is located at the point with coordinates (194m, 2.03 m), i.e. 0.33 meters over the horizontal line. Crosswind deflection is 0.696m.

If we input again the same data, and the value of range-wind 10m/s, the program gives a value of the departure angle of 0.1796 degree, and cross wind deflection is 0.667m.

Note that there is an insignificant difference of (-0.08) minutes in the angle of departure due to the range wind that blows in direction of shooting.

(b) The angle that the line of sight forms with the horizontal line that connects the target and the muzzle of the gun is

$$\alpha_h = -\frac{0.0381}{366} = -0.0001041 = -0.0059644°.$$

Using the data obtained in (a), we find that the angle of sight (the angle between the bullet line of departure and the line of sight) is

$$\alpha_s = \alpha_0 - \alpha_h = 0.18091 - (-0.0059644) = 0.18687° = 11.21 MOA.$$

(c) Inputting x-coordinate, 457 meters, and following the same procedure as in (a) and (b) we find:

Departure angle, 0.242676 degree; time of flight, 0.678 seconds; terminal speed, 533m/s, terminal angle, -0.3333 degree, trajectory vertex is located at the point with coordinates (246m, 2.26m), i.e. 0.56 meters over the horizontal line that connects the muzzle of the gun and the center of the target.

Angle of sight is 14.92 MOA.

Inclined Fire

(d) The y-coordinate of the target is

$$y_T = 366 \tan(30) + 1.7 = 213m.$$

Execute Program AngleM30.

Input: x-coordinate of target = 366; y-coordinate of target = 213; y-coordinate of muzzle = 1.7; initial speed = 853.44.

Output: Departure angle is 30.21875 degree. Distance to the target (inclined distance) is 422.74m.

Elementary calculations show that:
The **super elevation angle** is 0.21875 degree (13.13 MOA).
The **angle of sight** is 0.2247degree, (13.48 MOA).

Exercise 2.8

For the bullet of example 2.7:

(a) Find the departure angle needed to hit the target located at a horizontal distance of 366m from the shooter if the firearm muzzle is 1.6 meters above the sea level (the shooter is firing from a standing position); while the target is located at the sea level (y-coordinate of target is zero).

(b) Find the departure angle if the firing takes place in the same conditions as in (a) but 1200 meters over the sea level. The rifle is 1.6 meters over the terrain (y-coordinate of rifle is 1201.6.

(c) Find the departure angle needed to hit the target if the x-coordinate of target is 366m, the firing takes place 500 meters over the sea level, the muzzle of the gun is 1.6 meters over the ground (shooter is in standing position), while the target is 20 meters over the ground. Find the super elevation angle and the angle of sight.

(d) Find the departure angle needed to hit the target if the x-coordinate of target is 366m, the firing takes place the sea level, the muzzle of the gun and the target are 1.6 meters over the ground (shooter is in standing position). Find as well the angle of sight.

Answer: (a) -0.069393 degrees; (b) -0.074818 degrees; (c) 3.05713 degrees. (Note: Input x-coordinate of target, 366; y-coordinate of target, 520; y-coordinate of muzzle, 501.6; initial speed, 853.44). Super elevation angle is 0.182056 degree, and the angle of sight (angle between the line of sight and the firing line) is 0.1761 degree (10.57MOA). (d) 0.18091 degree (10.85 MOA), 0.1869 degree (11.21 MOA).

Example 2.9

For the bullet caliber 0.30 Ball M2 fired from a gun located at the altitude 120m over the sea level with initial speed 853.44m/s find the departure angle needed to hit a target located on a hill at a point with coordinates (400m, 250m). Find as well the time of flight, the speed and the angle at the target location.

Solution

Input: The coordinates of the target xo = 400m, yo =250m, the y-coordinate of the gun 120.
Output: Departure angle is18.1875 degree; time of flight is 0.61s; impact speed is 559m/s; impact angle is 17.71 Degree, inclined range is 420.54m.

Exercise 2.10

For the bullet of example 2.9 find the departure angle needed to hit the target located at the point with coordinates (400m, 1250m) if the location of the muzzle of the gun is in the altitude 1120 meter over the sea level.
Answer: 18.1875 degree, inclined distance to the target 420.60m.

Example 2.11 Downhill Shooting

For the bullet caliber 0.30 Ball M2 fired with a speed of 853.44m/s from a gun located on a hill at the altitude 250m over the sea level, find the departure angle needed to hit a target located at a point with coordinates (400m, 120m). Find as well the time of flight, the speed and the angle at the target location (see example 2.10, for comparison).

Solution

Input: The coordinates of the target xo = 400m, yo =120m, the y-coordinate of the gun 250, assuming that the origin of the coordinates is at the sea level.

Output: Departure angle is -17.841 degree; time of flight is 0.61s; impact speed is 562.4m/s; impact angle is -18.316 degree.

Exercise 2.12

Use the program AngleM30.Bas to find the departure angle needed to fire the bullet of example 2.11 to hit a target located at the point with coordinates (600m, 1800m) if the muzzle of the gun is located at the altitude 2000m.

Answer: Departure angle, -18.1274; inclined distance to target, 632.6m.

PC QuickBasic Program ANGLEM30.BAS
Standard Atmosphere, Wind is Present

'Given: Coordinates of the Target and the firearm,

      'Ballistics Coefficient—Function of Speed, initial speed

      'Find: Launching Angle, Time of Flight, Impact Speed, Impact Angle, etc

'_____

'CONTROL DATA

'Input: y-coordinate of target = 300 m, y-coordinate of target = 1.5

      'y-coordinate of muzzle = 0.25, projectile initial speed = 853.44.

'RESULTS: Departure Angle = 0.37964 degree, Time of Flight = 0.41 s

      'Terminal Speed = 633.5m/s, Terminal Angle = 0.0621 degree

'_____

'Note: Round the Input RANGE to the nearest 1; neglect the numbers after
decimal point

'_____

      'Functions and Sub. Prog.

```
DECLARE SUB y1z1v1w1 (x, y, z, v, w, y1, z1, v1, w1, koef, ys, yy, wind)
DECLARE SUB InfHyres (xx, n, koef, vv, vo, yo, xc1, cw, wind)
DECLARE SUB InfDales (x0, y0, z0, v0, w0, xm, ym, A, xc, yc, tc, vc, ac,
    vo, vv, xx, vo, cw)
DECLARE SUB NPxyzvw (nk, x, x0, y, y0, z, z0, v, v0, w, w0, h, h0, k, l, r, q)
DECLARE SUB NPkoef (k, l, r, q, h, y1, z1, v1, w1)
DECLARE SUB menu (cog, cof, xf, yf, xfu, yfu, t$)
DECLARE SUB y0z0 (y0, z0, A, vo, wind)
DECLARE SUB c (koef, y0, A)

'Variables
SCREEN 0
1 :
DIM m(4, 4), v(4)              'Intermediate values (k,l,r,q)
rendi = 4                'rend dif.
cog = 7: cof = 0
cikli = 0
A = 23                 'Initial Angle 23 degree
kendi = 22                 'd.Angle for maximum distance
kov = 1                'Test of the value of v0
tt = 1
```

```
'Solution
CLS

'Initial Data
menu cog, cof, 3, 10, 7, 70, "DATA INPUT"
InfHyres xx, n, koef, vv, vo, yo, xc1, cw, wind
hap = 1

IF wind = 0 THEN
gab = .01
ELSE
gab = 1
END IF

'Initial values
f:
x0 = 0: v0 = vv: w0 = 0
y0z0 y0, z0, A, vo, wind: h0 = hap
c koef, y0, A
ff:
FOR nk = 1 TO rendi
NPxyzvw nk, x, x0, y, y0, z, z0, v, v0, w, w0, h, h0, k, l, r, q
y1z1v1w1 x, y, z, v, w, y1, z1, v1, w1, koef, ys, yy, wind
NPkoef k, l, r, q, h, y1, z1, v1, w1
m(nk, 1) = k: m(nk, 2) = l
m(nk, 3) = r: m(nk, 4) = q
NEXT nk

'Estimations for new points
FOR i = 1 TO rendi
v(i) = 1 / 6 * (m(1, i) + 2 * m(2, i) + 2 * m(3, i) + m(4, i))
NEXT i

'New Points
x0 = x0 + h: y0 = y0 + v(1): z0 = z0 + v(2)
v0 = v0 + v(3): w0 = w0 + v(4)
xcc = x0 + wind * w0
IF ABS(xcc - xc1) <= .5 THEN
xc = xcc
```

```
yc = v0
tc = w0
ac = (180 / 3.141592654#) * ATN(z0)
vc = y0 / COS(ATN(z0))
END IF
xmm = x0 + w0 * wind
IF xmm > 30 AND ABS(z0) <= .00001 THEN
xm = xmm
ym = v0
END IF


'Tests the y-value
xT = x0 + wind * w0
IF kov = 1 THEN kov = -1: GOTO ff:
IF ABS(xT - xx) < gab AND v0 <= yo + (gab * TAN(A * 3.1415954# / 180))
    AND v0 >= yo + ((-1 * gab) * TAN(A * 3.1415954# / 180)) THEN
c:
'Display Results
CLS
PLAY "a8a16a32b8"
menu 12, 0, 5, 10, 11, 70, "RESULTS:"
COLOR 7
InfDales x0, y0, z0, v0, w0, xm, ym, A, xc, yc, tc, vc, ac, yo, vv, xx, vo, cw
CLS
GOTO 1:
END IF


IF ABS(xT - xx) < gab AND v0 > yo + (gab * TAN(A * 3.1415954# / 180)) THEN
t$ = "* ? *"
menu 18, 0, 10, 20, 14, 60, t$
COLOR 14
LOCATE 12, 30: PRINT "Wait a moment, Please (+)";
LOCATE 12, 53: PRINT tt
tt = tt + 1
COLOR 7
A = A - kendi
GOTO fff:
END IF
```

```
IF ABS(xT - xx) < gab AND v0 < yo + ((-1 * gab) * TAN(A * 3.1415954# / 180))
    THEN
t$ = "* ? *"
menu 18, 0, 10, 20, 14, 60, t$
COLOR 14
LOCATE 12, 30: PRINT "Wait a moment, Please (-)";
LOCATE 12, 53: PRINT tt
tt = tt + 1
COLOR 7
A = A + kendi
GOTO fff:
END IF
GOTO ff:
fff:
'Restart Cycle
cikli = cikli + 1
IF cikli = 20 THEN GOTO c:
kendi = kendi / 2
kov = 1
GOTO f:


SUB c (koef, y0, A)
koef = (.913405 - 9.944000000000001D-04 * y0 + .00000062# * y0 ^ 2) *
    6.3223913#
END SUB
SUB InfDales (x0, y0, z0, v0, w0, xm, ym, A, xc, yc, tc, vc, ac, yo, vv, xx, vo, cw)
aT = ATN(z0) * 180 / 3.141592654#
LOCATE 6, 16: PRINT "Launching Angle     = "; A; "degree"
LOCATE 7, 16: PRINT "Time of Flight      = "; INT((w0) * 1000 + .5) / 1000;
    "seconds";
LOCATE 8, 16: PRINT "Terminal Speed      = "; INT((y0 / COS(ATN(z0))) *
    100 + .5) / 100; "m/s"
LOCATE 9, 16: PRINT "Terminal Angle      = "; aT; "degree"
LOCATE 10, 16: PRINT "Trajectory Vertex = :"; "("; INT((xm) * 100 + .5) /
    100; ","; INT((ym) * 100 + .5) / 100; ")"
LOCATE 11, 16: PRINT "Cross-Wind Deflection    = "; INT((cw * (w0 - xx
    / (vo * COS(A * 3.14159265# / 180)))) * 1000 + .5) / 1000
```

```
LOCATE 13, 16: PRINT "Distance to the target   = "; INT(((xx ^ 2 + (vv - v0)
    ^ 2) ^ .5) * 100 + .05) / 100; "meter";
LOCATE 14, 16: PRINT "x, y coordinates of TARGET = "; "("; INT((xx) *
    100 + .5) / 100; ","; INT((v0) * 100 + .5) / 100; ")"
LOCATE 15, 16: PRINT "x,y coordinatees of GUN    = "; "("; 0; ","; vv; ")"
LOCATE 17, 18: PRINT "x-coordinate of a point[m] :"; INT((xc) * 100 + .5) / 100
LOCATE 18, 18: PRINT "Corresponding y [m]    :"; INT((yc) * 1000 + .5) / 1000
LOCATE 19, 18: PRINT "Corresponding Time [sec]  :"; INT((tc) * 100 + .5) / 100
LOCATE 20, 18: PRINT "Corresponding Speed [m/s] :"; INT((vc) * 100 + .5) / 100
LOCATE 21, 18: PRINT "Corresponding Angle [Deg]  :"; ac
LOCATE 22, 18: PRINT "Cross-Wind Deflection      :"; INT((cw * (tc - xc /
    (vo * COS(A * 3.14159265# / 180)))) * 1000 + .5) / 1000
COLOR 7
LOCATE 24, 11: PRINT "Pres [ P ] to repeat [ Esc ] to end  ";
cc$ = INPUT$(1)
IF cc$ = CHR$(27) THEN SCREEN 9: CLS : END
END SUB


SUB InfHyres (xx, n, koef, vv, vo, yo, xc1, cw, wind)
LOCATE 4, 13: INPUT "x-coordinate of TARGET [m]   ="; xx
LOCATE 5, 13: INPUT "y-coordinate of TARGET [m]   ="; yo
LOCATE 6, 13: INPUT "y-coord of GUN             ="; vv
LOCATE 7, 13: INPUT "Initial Speed [m/s]        ="; vo
LOCATE 9, 13: INPUT "X-coordinate of a point    ="; xc1
LOCATE 10, 13: INPUT "Range-Wind                = "; wind
LOCATE 11, 13: INPUT "Cross-Wind                = "; cw
CLS
END SUB
SUB menu (cog, cof, xf, yf, xfu, yfu, t$)
COLOR cog, cof
LOCATE xf - 1, yf: PRINT t$
LOCATE xf, yf: PRINT "É" + STRING$(yfu - yf, 205) + "»";
FOR i = xf + 1 TO xfu
LOCATE i, yf: PRINT "º" + SPACE$(yfu - yf) + "º";
NEXT
LOCATE xfu + 1, yf: PRINT "È" + STRING$(yfu - yf, 205) + "¼";
END SUB


SUB NPkoef (k, l, r, q, h, y1, z1, v1, w1)
```

```
k = h * y1: l = h * z1
r = h * v1: q = h * w1
END SUB

SUB NPxyzvw (nk, x, x0, y, y0, z, z0, v, v0, w, w0, h, h0, k, l, r, q)
IF nk = 1 THEN
x = x0: y = y0: z = z0
v = v0: w = w0: h = h0
GOTO fund:
END IF
IF nk = 2 OR nk = 3 THEN
x = x0 + (.5 * h): y = y0 + (.5 * k)
z = z0 + (.5 * l): v = v0 + (.5 * r)
w = w0 + (.5 * q)
GOTO fund:
END IF
IF nk = 4 THEN
x = x0 + h: y = y0 + k: z = z0 + l
v = v0 + r: w = w0 + q
END IF
fund:
END SUB

SUB y0z0 (y0, z0, A, vo, wind)
y0 = SQR(vo ^ 2 + wind ^ 2 - 2 * vo * wind * COS(A * 3.141592654# / 180))
y0 = y0 * COS(A * 3.141592654# / 180)
z0 = TAN(A * 3.141592654# / 180)
z0 = z0 / (1 - wind / (vo * COS(A * 3.141592654# / 180)))
END SUB

SUB y1z1v1w1 (x, y, z, v, w, y1, z1, v1, w1, koef, ys, yy, wind)
yy = y * SQR(1 + z ^ 2)
IF yy > 256! THEN
y1 = -1 * koef * ((289.08 - .006328 * v) / 289.08) ^ 4.4 * (yy - 240) / (3 * yy)
ELSE
y1 = -1 * koef * ((289.08 - .006328 * v) / 289.08) ^ 4.4 * .0001212 * yy ^ 2 / yy
END IF
z1 = -9.80665 / y ^ 2
v1 = z
w1 = 1 / y
END SUB
```

2.3 PC Program Angle122.Bas, Projectile 122mm

The PC program Angle122.Bas can be used to determine the departure angle of the projectile caliber 122mm of the Russian cannon Mod. 1960, fired with a speed of 885m/s, in order to hit a given a target at a known location. The ballistics coefficient (BC) of the projectile is a function of the departure angle.

The projectile trajectory is in the standard atmosphere, but the wind can be present during the flight.

The program will give wrong results if we input a range of fire that is greater than the maximum distance that the projectile can reach.

We assume that that the origin of the Cartesian system of coordinates is at the sea level.

The estimation of the departure angle using the PC program Angle122. Bas is done with satisfying accuracy, especially for the fire at the sea level. For firing in high mountains, the accuracy of the calculated angle of departure depends by the altitude of firing and the firing range.

Using Angle122.Bas

Example 2.13

A Russian cannon Mod.1960, fires a 122mm projectile with a speed of 885m/s. The atmosphere is standard.

(a) Find the departure angle needed to hit the target located at a horizontal range of 12,000m from the Russian cannon. Estimate as well the elements of the projectile flight at the trajectory point with abscissa 6000m. The cannon and the target are located at the sea level.

(b) Use the data of (a) to find the departure angle if the projectile is fired on a high mountain 1000 meters over the sea level. The cannon and the target are at the same altitude 1000m.

(c) Find the departure angle needed to hit the target located at a horizontal range of 12,000m from the Russian cannon. The cannon and the target are located at the sea level. There is a range wind 10m/s, and a crosswind 10m/s.

(d) Find the departure angle needed to hit the target located at the point with coordinates (12,000m, 500m) while the projectile is fired from the cannon located at the point with coordinates (0m, 100m). The projectile flight is in standard atmosphere, and in absence of wind.

Solution

(a) **Input**: x-coordinate of target, 12000; Initial speed, 885, x-coordinate of a point, 6000.

   Output: Elements of the trajectory at the horizontal range 12,000m are: Departure angle, 8.598755 degree; Time of flight, 22.11s; terminal speed, 365m/s; terminal angle, -15.451; trajectory vertex, (6909, 609).

   Elements of trajectory at the point with coordinate 6000m are: The corresponding y-coordinate, 596m; time, 8.55s, speed, 566m/s; angle, 1.7126.

(b) Altitude of fire, 1000m. Input the same data as in (a), but input for the y-coordinate of target and the y-coordinate of the cannon the value 1000.

   Elements of the trajectory at 12,000m, altitude 1000m are: Departure angle, 7.93366 degree; Time of flight, 20.98s; terminal speed, 395m/s; terminal angle, -13.5486; trajectory vertex, (6823, 546).

(c) The elements of trajectory in presence of wind are: The departure angle, 8.43356 degree; time of flight, 21.78s; terminal speed, 366m/s; terminal angle, -15.19; trajectory vertex, (6891, 596); cross-wind deflection, 81m.

(d) The departure angle is 10.54175 degree.

Example 2.14

(a) Find the departure angle needed to hit the target located in a horizontal range of 15,170m from the Russian cannon, Mod. 1960, if a 122mm projectile is fired with a speed of 885m/s.

(b) Estimate as well the elements of the projectile flight at the trajectory point with abscissa 7585m.

(c) Use the data in (a) to find the departure angle needed to hit the target, but if the fire is on a high mountain 1000 meters over the sea level.

Solution

(a) **Input**: x-coordinate of Target, 15170, initial speed 885, x-coordinate of a point in the trajectory 7585.

   Output: Departure angle is14.01953 degree; time of flight is 32.71s; impact speed is 316.37m/s; impact angle is -27.168 Degree; the abscissa of vertex is 8958m; the ordinate of vertex is 1363.74m.

(b) In the same way, we find that at the point of the trajectory with abscissa 7585, the projectile elements are: Altitude (ordinate) of the projectile is 1318.9m; time of flight is 11.85s; speed is 487.50m/s; the angle of flight is 3.62083 degree.

(c) Departure angle is 12.57739 degree; time 30.64s; speed, 335,5m/s; terminal angel, -23.647; trajectory vertex (8867, 2183).

Example 2.15

A Russian cannon Mod.1960, fires a 122mm projectile with a speed of 885m/s. The atmosphere is standard, and no wind is present. The cannon is located 100 meters over the sea level, while the target at the point with coordinates (9000m, 800m).

Find the departure angle needed to hit the given target.

Solution

Input: x-coordinate of target, 9000; y-coordinate of target, 800; the y-coordinate of the cannon, 100; initial speed, 885.

Output: departure angle, 9.71762 degree; the distance to the target is 9027.20m.

QBasic PC Program ANGLE122.BAS
Cannon 122mm only, Standard Atmosphere
'FIND: LAUNCHING Angle, Time of Flight, y-coordinate of a point, etc.
'GIVEN: Coordinates of target and the gun, Ballistics Coefficient, Launching Speed
'

            'CONTROL DATA
'INPUT: Range = 10,000: Launching Speed = 885:
            'x-coordinate of a point on Trajectory = 5500
'

'RESULTS: Launching Angle = 6.20104 [Degree], Time of Flight = 16.75 [Sec.]
            'Terminal Angle = -10.191 [Degree]
            'Coordinates of Trajectory vertex (5631, 347)
            'For x =5500m: y = 346.63m, Time = 7.62s,
            'Angle = 0.2119 Degree, Speed = 593.20m/s
'

'Note: Round the Input RANGE to the nearest 1; neglect the numbers after
    decimal point
'

            'Functions and Sub. Prog.
DECLARE SUB y1z1v1w1 (x, y, z, v, w, y1, z1, v1, w1, koef, ys, yy, wind)
DECLARE SUB InfHyres (xx, n, koef, vv, vo, yo, xc1, cw, wind)
DECLARE SUB InfDales (x0, y0, z0, v0, w0, xm, ym, A, xc, yc, tc, vc, ac,
    vo, vv, xx, vo, cw)
DECLARE SUB NPxyzvw (nk, x, x0, y, y0, z, z0, v, v0, w, w0, h, h0, k, l, r, q)
DECLARE SUB NPkoef (k, l, r, q, h, y1, z1, v1, w1)
DECLARE SUB menu (cog, cof, xf, yf, xfu, yfu, t$)
DECLARE SUB y0z0 (y0, z0, A, vo, wind)
DECLARE SUB c (koef, y0, A)

'Variables
SCREEN 0
1 :
DIM m(4, 4), v(4)   'Intermediate values (k,l,r,q)
rendi = 4   'rend dif.
cog = 7: cof = 0
cikli = 0
A = 23   'Initial Angle 23 degree
kendi = 22   'd.Angle for maximum distance
kov = 1   'Test of the value of v0

```
gab = 1
tt = 1

'Solution
CLS
'Initial Data
menu cog, cof, 3, 10, 7, 70, "DATA INPUT"
InfHyres xx, n, koef, vv, vo, yo, xc1, cw, wind
hap = 1

IF wind = 0 THEN
ELSE
gab = 1
END IF

'Initial values
f:
x0 = 0: v0 = vv: w0 = 0
y0z0 y0, z0, A, vo, wind: h0 = hap
c koef, y0, A
ff:
FOR nk = 1 TO rendi
NPxyzvw nk, x, x0, y, y0, z, z0, v, v0, w, w0, h, h0, k, l, r, q
y1z1v1w1 x, y, z, v, w, y1, z1, v1, w1, koef, ys, yy, wind
NPkoef k, l, r, q, h, y1, z1, v1, w1
m(nk, 1) = k: m(nk, 2) = l
m(nk, 3) = r: m(nk, 4) = q
NEXT nk

'Estimations for new points
FOR i = 1 TO rendi
v(i) = 1 / 6 * (m(1, i) + 2 * m(2, i) + 2 * m(3, i) + m(4, i))
NEXT i

'New Points
x0 = x0 + h: y0 = y0 + v(1): z0 = z0 + v(2)
v0 = v0 + v(3): w0 = w0 + v(4)

xcc = x0 + wind * w0
IF ABS(xcc - xc1) <= .5 THEN
```

```
xc = xcc
yc = v0
tc = w0
ac = (180 / 3.141592654#) * ATN(z0)
vc = y0 / COS(ATN(z0))
END IF

xmm = x0 + w0 * wind
IF xmm > 30 AND ABS(z0) <= .00001 THEN
xm = xmm
ym = v0
END IF

'Tests the y-value
xT = x0 + wind * w0
IF kov = 1 THEN kov = -1: GOTO ff:
IF ABS(xT - xx) < gab AND v0 <= yo + (gab * TAN(A * 3.1415954# / 180))
    AND v0 >= yo + ((-1 * gab) * TAN(A * 3.1415954# / 180)) THEN
c:
'Display Results
CLS
PLAY "a8a16a32b8"
menu 12, 0, 5, 10, 11, 70, "RESULTS:"
COLOR 7

InfDales x0, y0, z0, v0, w0, xm, ym, A, xc, yc, tc, vc, ac, yo, vv, xx, vo, cw
CLS
GOTO 1:
END IF

IF ABS(xT - xx) < gab AND v0 > yo + (gab * TAN(A * 3.1415954# / 180)) THEN
t$ = "* ? *"
menu 18, 0, 10, 20, 14, 60, t$
COLOR 14
LOCATE 12, 30: PRINT "Wait a moment, Please (+)";
LOCATE 12, 53: PRINT tt
tt = tt + 1
COLOR 7
A = A - kendi
```

```
GOTO fff:
END IF

IF ABS(xT - xx) < gab AND v0 < yo + ((-1 * gab) * TAN(A * 3.1415954# / 180))
    THEN
t$ = "* ? *"
menu 18, 0, 10, 20, 14, 60, t$
COLOR 14
LOCATE 12, 30: PRINT "Wait a moment, Please (-)";
LOCATE 12, 53: PRINT tt
tt = tt + 1
COLOR 7
A = A + kendi
GOTO fff:
END IF
GOTO ff:

fff:
'Restart Cycle
cikli = cikli + 1
IF cikli = 20 THEN GOTO c:
kendi = kendi / 2
kov = 1
GOTO f:

SUB c (koef, y0, A)
IF A >= 0 AND A <= 1.38333333# THEN koef = (1.82051 - 210.432 * (A *
    3.141592654# / 180) + 10066.5 * (A * 3.141592654# / 180) ^ 2 - 164950
    * (A * 3.141592654# / 180) ^ 3)'3200
IF A > 1.38333333# AND A <= 2.5666667# THEN koef = (.490432 - 14.05
    38 * (A * 3.141592654# / 180) + 298.238 * (A * 3.141592654# / 180) ^
    2 - 2328.38 * (A * 3.141592654# / 180) ^ 3) '5400
IF A > 2.5666667# AND A <= 4.8833333# THEN koef = (.352739 - 4.0664 * (A
    * 3.141592654# / 180) + 49.8947 * (A * 3.141592654# / 180) ^ 2 - 207.518
    * (A * 3.141592654# / 180) ^ 3) '8600
IF A > 4.8833333# AND A < 8.0066667# THEN koef = (.304165 - 1.65929 * (A
    * 3.141592654# / 180) + 13.5957 * (A * 3.141592654# / 180) ^ 2 - 35.3659
    * (A * 3.141592654# / 180) ^ 3) '11600
```

```
IF A > 8.0066667# AND A <= 12.9166667# THEN koef = (.324563 - 1.5842
    * (A * 3.141592654# / 180) + 9.56496 * (A * 3.141592654# / 180) ^
    2 - 17.802 * (A * 3.141592654# / 180) ^ 3) '14600
IF A > 12.9166667# AND A <= 18.3166667# THEN koef = (.158193 + .919
    854 * (A * 3.141592654# / 180) - 3.03889 * (A * 3.141592654# / 180) ^
    2 + 3.35924 * (A * 3.141592654# / 180) ^ 3) '17200
IF A > 18.3166667# AND A <= 45 THEN koef = (.251064 - .0064118# *
    (A * 3.141592654# / 180) + .0262451# * (A * 3.141592654# / 180) ^
    2 - .0064443# * (A * 3.141592654# / 180) ^ 3) '23800
END SUB

SUB InfDales (x0, y0, z0, v0, w0, xm, ym, A, xc, yc, tc, vc, ac, yo, vv, xx, vo, cw)
aT = ATN(z0) * 180 / 3.141592654#
LOCATE 6, 16: PRINT "Launching Angle    = "; A; "degree"
LOCATE 7, 16: PRINT "Time of Flight  = "; INT((w0) * 1000 + .5) / 1000; "seconds";
LOCATE 8, 16: PRINT "Terminal Speed    = "; INT((y0 / COS(ATN(z0))) *
    100 + .5) / 100; "m/s"
LOCATE 9, 16: PRINT "Terminal Angle    = "; aT; "degree"
LOCATE 10, 16: PRINT "Trajectory Vertex = :"; "("; INT((xm) * 100 + .5) /
    100; ","; INT((ym) * 100 + .5) / 100; ")"
LOCATE 11, 16: PRINT "Cross-Wind Deflection     = "; INT((cw * (w0 - xx
    / (vo * COS(A * 3.14159265# / 180)))) * 1000 + .5) / 1000
LOCATE 13, 16: PRINT "Distance to the target  = "; INT((((xx ^ 2 + (vv - v0)
    ^ 2) ^ .5) * 100 + .05) / 100; "meter";
LOCATE 14, 16: PRINT "x, y coordinates of TARGET = "; "("; INT((xx) *
    100 + .5) / 100; ","; INT((v0) * 100 + .5) / 100; ")"
LOCATE 15, 16: PRINT "x,y coordinatees of GUN    = "; "("; 0; ","; vv; ")"
LOCATE 17, 18: PRINT "x-coordinate of a point[m] :"; INT((xc) * 100 + .5) / 100
LOCATE 18, 18: PRINT "Corresponding y [m]      :"; INT((yc) * 1000 + .5) / 1000
LOCATE 19, 18: PRINT "Corresponding Time [sec]  :"; INT((tc) * 100 + .5) / 100
LOCATE 20, 18: PRINT "Corresponding Speed [m/s] :"; INT((vc) * 100 + .5) / 100
LOCATE 21, 18: PRINT "Corresponding Angle [Deg]   :"; ac
LOCATE 22, 18: PRINT "Cross-Wind Deflection      :"; INT((cw * (tc - xc /
    (vo * COS(A * 3.14159265# / 180)))) * 1000 + .5) / 1000
COLOR 7
LOCATE 24, 11: PRINT "Pres [ P ] to repeat [ Esc ] to end";
cc$ = INPUT$(1)
IF cc$ = CHR$(27) THEN SCREEN 9: CLS : END
```

```
END SUB

SUB InfHyres (xx, n, koef, vv, vo, yo, xc1, cw, wind)
LOCATE 4, 13: INPUT "x-coordinate of TARGET [m]    ="; xx
LOCATE 5, 13: INPUT "y-coordinate of TARGET [m]    ="; yo
LOCATE 6, 13: INPUT "y-coord of GUN            ="; vv
LOCATE 7, 13: INPUT "Initial Speed [m/s]         ="; vo
LOCATE 9, 13: INPUT "X-coordinate of a point     ="; xc1
LOCATE 10, 13: INPUT "Range-Wind            = "; wind
LOCATE 11, 13: INPUT "Cross-Wind            = "; cw
CLS
END SUB

SUB menu (cog, cof, xf, yf, xfu, yfu, t$)
COLOR cog, cof
LOCATE xf - 1, yf: PRINT t$
LOCATE xf, yf: PRINT "É" + STRING$(yfu - yf, 205) + "»";
FOR i = xf + 1 TO xfu
LOCATE i, yf: PRINT "º" + SPACE$(yfu - yf) + "º";
NEXT
LOCATE xfu + 1, yf: PRINT "È" + STRING$(yfu - yf, 205) + "¼";
END SUB

SUB NPkoef (k, l, r, q, h, y1, z1, v1, w1)
k = h * y1: l = h * z1
r = h * v1: q = h * w1
END SUB

SUB NPxyzvw (nk, x, x0, y, y0, z, z0, v, v0, w, w0, h, h0, k, l, r, q)
IF nk = 1 THEN
x = x0: y = y0: z = z0
v = v0: w = w0: h = h0
GOTO fund:
END IF
```

```
IF nk = 2 OR nk = 3 THEN
x = x0 + (.5 * h): y = y0 + (.5 * k)
z = z0 + (.5 * l): v = v0 + (.5 * r)
w = w0 + (.5 * q)
GOTO fund:
END IF

IF nk = 4 THEN
x = x0 + h: y = y0 + k: z = z0 + l
v = v0 + r: w = w0 + q
END IF
fund:
END SUB

SUB y0z0 (y0, z0, A, vo, wind)
y0 = SQR(vo ^ 2 + wind ^ 2 - 2 * vo * wind * COS(A * 3.141592654# / 180))
y0 = y0 * COS(A * 3.141592654# / 180)
z0 = TAN(A * 3.141592654# / 180)
z0 = z0 / (1 - wind / (vo * COS(A * 3.141592654# / 180)))
END SUB

SUB y1z1v1w1 (x, y, z, v, w, y1, z1, v1, w1, koef, ys, yy, wind)
yy = y * SQR(1 + z ^ 2)
IF yy > 256! THEN
y1 = -1 * koef * ((289.08 - .006328 * v) / 289.08) ^ 4.4 * (yy - 240) / (3 * yy)
ELSE
y1 = -1 * koef * ((289.08 - .006328 * v) / 289.08) ^ 4.4 * .0001212 * yy ^
    2 / yy
END IF
z1 = -9.80665 / y ^ 2
v1 = z
w1 = 1 / y
END SUB
```

2.4 PC Program ANMECO.BAS, Non Standard Atmosphere

The PC program Anmeco.Bas can be used to determine the departure angle of a projectile of any firearm, when the projectile flight is in a non-standard atmosphere and in the presence of wind.

The initial speed of the projectile, the departure angle, the ballistics coefficient, as well as the coordinates of the target and the firearm are known.

The program can be used when the projectile characteristics are standard or non-standard. For example, when it is a small change in the departure speed, or a small change in the projectile mass.

Anmeco.Bas can be used for uphill shooting as well. The program can be modified to be used for the downhill shooting as well.

The PC program can be employed instead of the PC program AngleC. Bas, already presented in section 2.1.

Use of Anmeco.Bas

Example 2.16 Non Standard Atmosphere

Use the PC program "Anmeco.Bas" to find the departure angle needed to hit a target located at the sea level at a horizontal range of 500m if a 7.62mm Russian bullet is fired with a speed of 735m/s. The BC of the bullet is 4.116 (form factor, .56).

(a) The projectile flight is in non-standard atmosphere, while the projectile characteristics are standard: Temperature of air is 30 degree Celsius; pressure is 765mm; Hg. pressure of water vapor is 6.35 mm Hg. range wind is 5 m/s; crosswind is 5m/s.

Projectile mass is standard 0.0079kg; powder load temperature is 30 degree Celsius, projectile speed is 735m/s.

(b) Solve the same problem presented in (a), but the fire is in winter time: Temperature of air is -5 degree Celsius; pressure is 745mm; Hg., pressure of water vapor is 6.35 mm Hg.; range wind is 5 m/s.

Projectile mass is standard 0.0079kg; powder load temperature is -5 degree Celsius, projectile speed is 735m/s.

(c) What is the departure angle if the shooting is in standard atmosphere?

Solution

(a) **Input**: x-coordinate of target, 500; initial speed, 735; air temperature, 30; powder load temperature, 30; pressure, 765; pressure of water vapor, 6.35; range wind 5; cross wind, 5; ballistics coefficient, 4.116.

　　Output: Departure angle, 0.4038 degree; time, 0.94; terminal speed, 384; terminal angle, -0.634; vertex (279, 1.09); cross wind deflection, 1.3m.

(b) In the same way as in (a) we find:

　　Departure angle, 0.460205 degree; time, 1.01; terminal speed, 349; terminal angle, -0.747; vertex (280, 1.26); cross wind deflection, 1.69m.

(c) Departure angle is 0.43066.

Remark

Comparing the results obtained in (a), (b), and (c) we see that there is a significant change in the departure angle, around 3.70 MOA.

Exercise 2.17

A 7.62 x 51 mm NATO (.308 Winchester) cartridge loaded with Sierra's 30 caliber 168 grain Hollow Point Boat Tail Match bullet has an initial speed of 807.72m/s. The Siacci ballistics coefficient of the given bullet is 2.886.

(a) Use "Anmeco.Bas" to find the departure angle needed to hit a target located at the point with coordinates (500m, 1000m). The rifle is at an altitude of 950m over the sea level. The atmosphere is non-standard.

　　At the sea level the temperature of air is 0 degree Celsius, the atmospheric pressure is 740 mm; the pressure of water vapor is 6.35mm; the projectile mass is 0.0109kg; the powder load temperature is 0 degree Celsius; the range-wind is 10m/s, in the opposite direction of fire; the cross wind is 4m/s.

(b) Find the departure angle if the rifle is at the sea level, the target is located at the point with coordinates (500m, 50m). The atmosphere is not standard:

Temperature of air is 0 degree Celsius; the atmospheric pressure is 740 mm; the pressure of water vapor is 6.35mm; the projectile mass is 0.0109kg; the powder load temperature is 0 degree Celsius.

The range wind (in the positive direction of fire) is 10 m/s, cross wind is 4m/s.

(c) Find the departure angle if the rifle is at the sea level, the target is located at the point with coordinates (500m, 50m), but the atmosphere at the sea level is standard: Temperature of air is 15 degree Celsius; the atmospheric pressure is 750 mm; the pressure of water vapor is 6.35mm; the projectile mass is 0.0109kg; the powder load temperature is 15 degree Celsius.

There is a range wind in the direction of fire is 10 m/s, cross wind is 4m/s.

Answer: (a) Departure angle is 6.08106 degree; time is 0.781s, speed is 518m/s; terminal angle is 5.316457 degree; firing distance is 502.50 meters; cross wind deflection, 0.632.

(b) Departure angle is 6.00586 degree; time is 0.79s, speed is 506m/s; impact angle is 5.3028 degree; firing distance is 502.5 meters; cross wind deflection, 0.7.

(c) Departure angle is 6.02734 degree; time is 0.78s, speed is 503m/s; impact angle is 5.398 degree; cross wind deflection, 0.64m.

Remark

Comparing the results obtained in (a), (b), and (c) we see that there are some small changes in the departure angle (lees than or equal to 1.3 MOA).

Exercise 2.18 Cannon 76 mm

The departure speed of a projectile fired by a 76.2mm cannon is 588m/s.

(a) Find the departure angle needed to hit a target at 5060m from the cannon. The target and the cannon are at the sea level. The atmosphere is standard. The ballistics coefficient is 0.7045.
(b) Find the departure angle if there is a 10 degree increase in the temperature of air, but all other parameters are as in (a).
(c) Find the departure angle if there is a wind of 10m/s in the direction of flight, but all other parameters are as in (a).

Answer: (a) Departure angle is 10 degree. (b) Departure angle is 9.6877 degree. (c) 9.58838 degrees.

PC Program ANMECO.Bas
Non Standard Atmosphere
'Find:　Departure Angle, Time of Flight, Impact Speed, Terminal Angle.
'Given: The coordinates of the target and the location of the muzzle of the cannon
　　　'& Ballistics Coefficient as Function of Angle;
　　　'Temperature of air and thrusting load of projectile are known;
　　　'The weight of projectile and the air humidity are known
　　　'Projectile flight is in presence of wind.
'

'CONTROL DATA
'INPUT:　x0 = 10,000m; projectile speed = 885;
　　　　'Temperature of air = 15, Temperature of propellant= 15
　　　　'Pressure = 750; Pressure of Water vapor = 6.35; projectile mass = 27.3;
　　　　'change in mass = 0; range wind = 10, cross wind, 0.
'

'RESULTS:　Launching Angle = 6.12 Degree, Time of Flight = 16.6s
　　　　'Impact Speed = 419m/s, Impact Angle = -10.106 Degree
　　　　'Coordinates of the trajectory vertex (5610, 339)
'

'Functions and Sub. Prog.
DECLARE SUB y1z1v1w1 (x, y, z, v, w, y1, z1, v1, w1, koef, pa1, wind, ys, yy, pa, ta1)
DECLARE SUB InfHyres (xx, voo, vo1, ta, ta1, pa, pa1, ea, m, dm, tp, wind,
　　　xc, yc, vo, yc1, cw, xx1, koef)
DECLARE SUB InfDales (x0, y0, z0, v0, w0, xc, yc, A, aT, vo, yc1, xm, ym,
　　　cw, xx, voo, xa, ya, ta, va, aa)
DECLARE SUB NPxyzvw (nk, x, x0, y, y0, z, z0, v, v0, w, w0, h, h0, k, l, r, q)
DECLARE SUB NPkoef (k, l, r, q, h, y1, z1, v1, w1)
DECLARE SUB menu (cog, cof, xf, yf, xfu, yfu, t$)
DECLARE SUB y0z0 (y0, z0, A, vo1, wind)
DECLARE SUB c (koef, m, dm)
'Variables
SCREEN 0
1 :
DIM m(4, 4), v(4)　　　　　'Intermediate values (k,l,r,q)
rendi = 4　　　　　　'rend dif.
cog = 7: cof = 0
cikli = 0
A = 23　　　　　　　'Initial Angle 23 degree
kendi = 22　　　　　　'Angle [Degree] for maximum distance
kov = 1　　　　　　　'Test of the value of v0

```
tt = 1

'Solution
CLS
'Initial Data
menu cog, cof, 3, 10, 7, 70, "DATA INPUT"
InfHyres xx, voo, vo1, ta, ta1, pa, pa1, ea, m, dm, tp, wind, xc, yc, yc1, vo, cw, xx1, koef
hap = 1

IF wind = 0 THEN
gab = .01
ELSE
gab = 1
END IF

'Initial values
f:
x0 = 0: v0 = vo: w0 = 0
y0z0 y0, z0, A, vo1, wind: h0 = hap
c koef, m, dm
ff:
FOR nk = 1 TO rendi
NPxyzvw nk, x, x0, y, y0, z, z0, v, v0, w, w0, h, h0, k, l, r, q
y1z1v1w1 x, y, z, v, w, y1, z1, v1, w1, koef, pa1, wind, ys, yy, pa, ta1
NPkoef k, l, r, q, h, y1, z1, v1, w1
m(nk, 1) = k: m(nk, 2) = l
m(nk, 3) = r: m(nk, 4) = q
NEXT nk

'Estimation for new points
FOR i = 1 TO rendi
v(i) = 1 / 6 * (m(1, i) + 2 * m(2, i) + 2 * m(3, i) + m(4, i))
NEXT i

'New Points
x0 = x0 + h: y0 = y0 + v(1): z0 = z0 + v(2)
v0 = v0 + v(3): w0 = w0 + v(4)

xmm = x0 + wind * w0
IF ABS(z0) <= .0001 THEN
```

```
xm = xmm
ym = v0
END IF


xaa = x0 + wind * w0
IF ABS(xx1 - xaa) <= 1 THEN
xa = xaa
ya = v0
ta = w0
va = y0 / COS(ATN(z0))
aa = (180 / 3.1415169542#) * ATN(z0)
END IF


'Tests the y-value
xT = x0 + wind * w0
IF kov = 1 THEN kov = -1: GOTO ff:
IF ABS(xT - xc) < gab AND ABS(v0 - yc) <= (gab * TAN(A * 3.1415954# / 180))
    THEN
c:
'DISPLAY of RESULTS
CLS
PLAY "a8a16a32b8"
menu 12, 0, 5, 10, 11, 70, "RESULTS:"
COLOR 7
InfDales x0, y0, z0, v0, w0, A, xc, yc, aT, vo, yc1, xm, ym, cw, xx, voo, xa, ya, ta, va, aa
CLS
GOTO 1:
END IF
IF ABS(xT - xc) < gab AND (v0 - yc) > (gab * TAN(A * 3.1415954# / 180)) THEN
t$ = "* ? *"
menu 18, 0, 10, 20, 14, 60, t$
COLOR 14
LOCATE 12, 30: PRINT "Wait a moment, Please (+)";
LOCATE 12, 53: PRINT tt
tt = tt + 1
COLOR 7
A = A - kendi
GOTO fff:
END IF
```

```
IF ABS(xT - xc) < gab AND (v0 - yc) < ((-1 * gab) * TAN(A * 3.1415954# / 180))
    THEN
t$ = "* ? *"
menu 18, 0, 10, 20, 14, 60, t$
COLOR 14
LOCATE 12, 30: PRINT "Wait a moment, Please (-)";
LOCATE 12, 53: PRINT tt
tt = tt + 1
COLOR 7
A = A + kendi
GOTO fff:
END IF
GOTO ff:

fff:
'Restart Cycle
cikli = cikli + 1
IF cikli = 20 THEN GOTO c:
kendi = kendi / 2
kov = 1
GOTO f:

SUB c (koef, m, dm)
koef = koef * (1 - dm / m)
END SUB

SUB InfDales (x0, y0, z0, v0, w0, A, xc, aT, yc, vo, yc1, xm, ym, cw, xx, voo,
    xa, ya, ta, va, aa)
aT = ATN(z0) * 180 / 3.141592654#
LOCATE 6, 16: PRINT "Departure Angle          :"; A; "Degree"
LOCATE 7, 16: PRINT "Time of Flight to Target  :"; INT((w0) * 1000 + .5) / 1000
LOCATE 8, 16: PRINT "Terminal Speed            :"; INT((y0 / COS(ATN(z0)))
    * 100 + .5) / 100; "m/s"
LOCATE 9, 16: PRINT "Terminal Angle            :"; aT; "degree"
LOCATE 10, 16: PRINT "Coordinates of Vertex     :"; "("; INT((xm) * 100 +
    .5) / 100; ","; INT((ym) * 100 + .5) / 100; ")"
LOCATE 11, 16: PRINT "Cross-Wind Deflection     :"; INT((cw * (w0 - xx /
    (voo * COS(A * 3.14159265# / 180)))) * 1000 + .5) / 1000
```

```
LOCATE 13, 16: PRINT "Location of TARGET        :"; "("; xx; yc1; ")"
LOCATE 14, 16: PRINT "Location of FIREARM       :"; "(0,"; vo; ")"
LOCATE 15, 16: PRINT "Distance to TARGET      :"; (xx ^ 2 + (yc1 - vo) ^ 2) ^ .5
LOCATE 17, 16: PRINT "X-Coordinate of the Point  :"; INT((xa) * 100 + .5) / 100
LOCATE 18, 16: PRINT "Y-Coordinate of the Point  :"; INT((ya) * 100 + .5) / 100
LOCATE 19, 16: PRINT "Time of Flight to the Point :"; INT((ta) * 100 + .5) / 100
LOCATE 20, 16: PRINT "Speed at the Point         :"; INT((va) * 100 + .5) / 100
LOCATE 21, 16: PRINT "Angle at the Point          :"; aa
LOCATE 22, 16: PRINT "Cross Wind Deflection        :"; INT((cw * (ta - xa /
    (voo * COS(A * 3.14159265# / 180)))) * 1000 + .5) / 1000
COLOR 7
LOCATE 24, 11: PRINT "Pres [ P ] to repeat [ Esc ] to end ";
cc$ = INPUT$(1)
IF cc$ = CHR$(27) THEN SCREEN 9: CLS : END
END SUB


SUB InfHyres (xx, voo, vo1, ta, ta1, pa, pa1, ea, m, dm, tp, wind, xc, yc, yc1,
    vo, cw, xx1, koef)
LOCATE 4, 13: INPUT "x-coordinate of TARGET [m]    = "; xx
xc = xx
LOCATE 5, 13: INPUT "y-Coordinate of TARGET [m]    = "; yc
yc1 = yc
LOCATE 6, 13: INPUT "y-Coordinate of FIREARM [m]   = "; vo
LOCATE 7, 13: INPUT "Departure Speed [m/s]        = "; voo
LOCATE 9, 13: INPUT "Temperature of Air [C]        = "; ta
LOCATE 10, 13: INPUT "Propellant Temperature [C]   = "; tp
LOCATE 11, 13: INPUT "Atmospheric Pressure [mm Hg] = "; pa
LOCATE 12, 13: INPUT "Pressure of Water vapor [mm Hg]= "; ea
LOCATE 13, 13: INPUT "Projectile Standard Mass [kg] = "; m
LOCATE 14, 13: INPUT "Change in Projectile Mass    = "; dm
LOCATE 15, 13: INPUT "Range-Wind [m/s]             = "; wind
LOCATE 16, 13: INPUT "Cross-Wind [m/s]             = "; cw
LOCATE 17, 13: INPUT "X-Coordinate of a Point [m]  = "; xx1
LOCATE 18, 13: INPUT "Ballistics Coefficient BC    = "; koef
CLS
ta = ta + 273.15
pa1 = ta / (1 - .3785 * ea / pa)
vo1 = (voo - .4 * voo * (dm / m) + .00125 * voo * (tp - 15))
END SUB
```

```
SUB menu (cog, cof, xf, yf, xfu, yfu, t$)
COLOR cog, cof
LOCATE xf - 1, yf: PRINT t$
LOCATE xf, yf: PRINT "É" + STRING$(yfu - yf, 205) + "»";
FOR i = xf + 1 TO xfu
LOCATE i, yf: PRINT "º" + SPACE$(yfu - yf) + "º";
NEXT
LOCATE xfu + 1, yf: PRINT "È" + STRING$(yfu - yf, 205) + "¼";
END SUB


SUB NPkoef (k, l, r, q, h, y1, z1, v1, w1)
k = h * y1: l = h * z1
r = h * v1: q = h * w1
END SUB


SUB NPxyzvw (nk, x, x0, y, y0, z, z0, v, v0, w, w0, h, h0, k, l, r, q)
IF nk = 1 THEN
x = x0: y = y0: z = z0
v = v0: w = w0: h = h0
GOTO fund:
END IF
IF nk = 2 OR nk = 3 THEN
x = x0 + (.5 * h): y = y0 + (.5 * k)
z = z0 + (.5 * l): v = v0 + (.5 * r)
w = w0 + (.5 * q)
GOTO fund:
END IF

IF nk = 4 THEN
x = x0 + h: y = y0 + k: z = z0 + l
v = v0 + r: w = w0 + q
END IF
fund:
END SUB


SUB y0z0 (y0, z0, A, vo1, wind)
y0 = SQR(vo1 ^ 2 + wind ^ 2 - 2 * vo1 * wind * COS(A * 3.141592654# / 180))
y0 = y0 * COS(A * 3.141592654# / 180)
```

```
z0 = TAN(A * 3.141592654# / 180)
z0 = z0 / (1 - wind / (vo1 * COS(A * 3.141592654# / 180)))
END SUB

SUB y1z1v1w1 (x, y, z, v, w, y1, z1, v1, w1, koef, pa1, wind, ys, yy, pa, ta1)
yy = y * SQR(1 + z ^ 2)
ta1 = (289.08 / pa1) ^ .5
IF (yy * ta1) >= 256 THEN
y1 = -1 * koef * (pa / 750) * (((pa1 - .006328 * v) / pa1) ^ 4.4) * (ta1 *
    yy - 240) / (3 * yy)
ELSE
y1 = -1 * 1.212 * 10 ^ - 4 * koef * (pa / 750) * (((pa1 - .006328 * v) / pa1) ^
    4.4) * (ta1 * yy) ^ 2 / (y * SQR(1 + z ^ 2))
END IF
z1 = -9.80665 / y ^ 2
v1 = z
w1 = 1 / y
END SUB
```

2.5 PC Program Angmet.Bas, Cannon 122mm

The PC program Angmet.Bas can be used to determine the departure angle of a projectile of the 122mm Russian cannon Mod.1960, when the projectile flight is in non-standard atmosphere and in presence of range wind. The program can be also used when the projectile characteristics are standard.

The coordinates of the target and the cannon are known.

Angmet.Bas is a good substitute for the range tables of 122mm Russian cannon, Mod. 1960, for the projectiles launched with speed 885m/s at the sea level, or close to it.

The PC program Angmet.Bas can be used instead of the PC programs Angle122.Bas considering that the projectile flight is in standard atmosphere, and that the projectile characteristics are standard.

The PC program Anta122.Bas (Ref. Exterior Ballistics with Applications" is not present in this book, since it is substituted by the (modified) program Angmet.Bas.

Angmet.Bas can be used for shooting when the target and the cannon are at the same altitude over the sea level. The results are approximate. The accuracy is better the lower altitude is, or the smaller the range is.

The PC program Angmet.Bas can be used for inclined shooting as well. The results obtained for inclined shooting are approximate. The accuracy is greater the closer is the altitude of the cannon to the sea level.

Angmet.Bas can be sued to estimate the corrections in the projectile range caused by small deviations of the characteristics of the atmospheric and the projectile from the standard values (see example 2.24).

Note. The use of the PC program Angmet.Bas (as well as Angle122. Bas) for small charges of the powder load, i.e. when the projectile speed is 770m/s, 660m/s or 550m/s is not recommended. The results are approximate for the initial speed of 770m/s, but not accurate enough for the practice of firing especially for large ranges.

Anyway, those PC Programs can be modified for such departure speeds of a 122mm projectile.

Use of Angmet.Bas

Example 2.19 Standard Atmosphere
Projectile 122mm, Russian cannon Mod.1960

Find the departure angle needed to hit a target located at a horizontal range of 12,400m from the cannon if a 122mm projectile is fired from the Russian cannon Mod.1960 with a speed of 885m/s. Estimate as well the elements of the projectile flight at the trajectory point with abscissa 5150m.

The flight is in standard atmosphere and the projectile characteristics are standard, i.e. the temperature is 15 degree Celsius, the pressure is 750mm. Hg., the pressure of water vapor 6.35mm Hg., the projectile mass is 27.3kg, the powder load temperature is 15 degree Celsius, the range wind speed and the cross wind speed are 0.

Find as well the other elements at the point of impact, and the trajectory vertex.

Solution

Input: Range = 12,400: Departure Speed = 885; Air Temperature = 15; Atmospheric Pressure = 750; Pressure of Water Vapor = 6.35; Projectile Mass = 27.30; Change in Projectile mass = 0; Powder load (black powder) temperature = 15; Coordinate x of a point = 5150.

We assume that that the origin of the coordinates is at the sea level.

Output: Range 12400m: Departure angle is 9.165405 degree; time is 23.31s; Impact speed is 356.40m/s; Impact angle is -16.720degree; Vertex (7171, 679), For the point on the trajectory with abscissa 5150 meters, the program output is: The corresponding ordinate y is 611m; time of flight 7.11s; projectile speed is 604m/s; angle is 3.66 degree.

Exercise 2.20 Non Standard Atmosphere

Find the departure angle needed to hit the target located at a horizontal range of 12,400m if the projectile of 122mm Russian cannon is fired with a speed of 885m/s. The flight is in non-standard atmosphere, wind is present, i.e.:

The temperature is 20 degree Celsius, pressure is 760mm Hg. pressure of water vapor 7.2mm Hg. projectile mass is 27.15kg, powder load temperature is 20 degree Celsius, the range wind speed is -10m/s. There is also a cross wind with a speed of 5m/s. Find as well the trajectory vertex, and the cross wind deflection.

Answer: Departure angle, 9.1555 degree; vertex, (7168, 686); cross deflection, 46.11m.

Example 2.21 Standard Atmosphere, Verifying Calculations
Projectile 122mm, Russian cannon Mod.1960

Find the departure angle needed to hit a target located in a horizontal range of 13,200 meter if a projectile of a Russian122mm cannon is fired with a speed of 885m/s.

The projectile flight is in standard atmosphere, and the projectile characteristics are standard, i.e. temperature of air is 15 degree Celsius, pressure is 750 mm Hg, pressure of water vapor is 6.35 mm Hg, wind speed is zero.

Projectile mass is 27.3 kg, change in projectile mass is 0; powder load temperature is 15 degree Celsius. Wind is not present.

Find as well all the projectile elements at the impact point and compare the obtained data with the data given in the Russian range table of the cannon:

Range Table, Range 13,200m

Departure angle is 10.417 degree; time of flight is 26.00s, impact speed is 332.00m/s; impact angle -20.00 degree; the altitude of maximum height is 843m.

In the range table of cannon 122mm there are not displayed data on the x-coordinate of the trajectory vertex.

Solution

Input: x-coordinate of target, 13200; y-coordinate of target, 0; y-coordinate of cannon, 0; projectile speed, 885; temperature of air, 15; temperature of powder load, 15; pressure, 750; pressure of water vapor, 6.35; projectile mass = 27.3; change projectile mass, 0; range wind, 0.

Output: For the range 13200m: Angle is 10.4207 degree; time is 25.86s; speed is 340.4; impact angle, -19.535; trajectory vertex is located at (7693m, 840m).

Remark

Comparing the calculated data with the data of the range table of 122mm cannon, we see that the results obtained using **Angmet.Bas** are accurate.

Exercise 2.22 Standard Atmosphere
Projectile 122mm, Russian cannon Mod.1960

(a) Find the departure angle needed to hit a target located at the point with coordinates (14400 m, 1000m). The cannon is at the same altitude as the target, 1000m over the sea level. Projectile initial speed is 885m/s.

Projectile moves in a standard atmosphere: At the sea level, the temperature of air is 15 degree Celsius; the atmospheric pressure is 750 mm; the pressure of water vapor is 6.35mm; the projectile mass is 27.30; the powder load temperature is 15 degree Celsius; there is no wind.

Find as well the coordinates of the trajectory vertex.

(b) Solve the same problem when the projectile range is 9200m.

Answer

(a) Range 14,400m:

Departure Angle is 11.295 degree; time is 28.1s, speed is 345.73m/s; terminal angle is -20.91m/s; trajectory vertex, (8374, 1991), i.e. the trajectory vertex is 991 meter high.

Note: The range table of the 122mm cannon for the range 14400 meters gives the departure angle 11.1 degree.

(b) Range 9200m:

Departure Angle is 5.13171 degree; time is 14.33s; speed is 475.4m/s; impact angle is -7.763 degree; trajectory vertex, (5084m, 1253m), trajectory height 253m over the horizontal line at altitude 1000 meters.

Note: The range table of the 122mm cannon shows that for the range 9200m the departure angle is 5.41667 degree.

Exercise 2.23 Non Standard Atmosphere

Use the program Angmet.Bas to find the departure angle needed to hit a target located at the point with coordinates (14150 m, 800m) while the cannon is at an altitude 400m over the sea level. Projectile initial speed is 882m/s.

The atmosphere is non-standard; at the sea level: Temperature of air is 5 degree Celsius; atmospheric pressure is 740 mm; pressure of water vapor is 7.2mm; projectile mass is 27.15 (projectile standard mass is 27.30 kg); powder load temperatures is 10 degree Celsius; range wind is -10m/s (against the projectile flight). Find as well the correction in the direction of the z-axis, if there is a 3 o'clock cross wind of 8m/s.

Find also the coordinates of the trajectory vertex.

Note. The standard mass of 122mm projectile is 27.3 g. There is a change in the mass of projectile equal to 27.15-27.30 = -0.15kg.

Remark

Answer: 13.9096 degree; vertex (8852, 1746); cross wind deflection is 104m (the direction of fire should be corrected rotating the firearm clock-wise by an angle of

$$\frac{104}{14150} \cdot \frac{10,800}{\pi} = 25.27 MOA .$$

Example 2.24 Corrections

A Russian cannon Mod.1960 fires a 122mm projectile. Use the PC program Angmet.Bas to find the departure angle corrections for the following deviations of firing data from the standard values, if the horizontal range is 12000.

- Changes in Atmospheric Characteristics
  Change in temperature the of air +10 degree Celsius; increase in atmospheric pressure, +10mm Hg.; change in range wind, -10m/s (the range wind is opposite to the positive direction of flight); change in cross wind, 10m/s.
- Change in Projectile Characteristics
  Change in the powder load temperature, +10 degree Celsius; increase in projectile mass, 0.67% (0.18kg); decrease in the departure speed, -1% (-8.85 m/s).

Solution

We will estimate the changes in angle for each deviation of the above parameters from the standard ones (see as well: the example 2, page 309-314, "Exterior Ballistics with Applications").

First, we find the departure angle in standard atmosphere, assuming the projectile characteristics at the sea level are standard.

Inputting in the PC program Angmet.Bas the standard data:

Temperature 15 degree Celsius; pressure, 750mm Hg.; pressure of water vapor, 6.35mm Hg.; the range wind and the cross wind are zero, projectile mass, 27.30 kg; change in projectile mass is zero; temperature of the thrusting charge of the projectile is 15 degree Celsius, we find that the departure angle is 8.598755 degree.

To estimate the change in range because of an increase of 10 degrees in temperature of air we input the air temperature 25 degree Celsius and find that the departure angle is 8.401031.

The change in the value of the departure angle, because of a (+10) degree change in air temperature, is

$$\Delta\alpha = 8.401367 - 8.598755 = 0.1974° = -11.84 MOA \, .$$

In the same way, we find the change in the departure angle as result of a change of +10 mm Hg, in the pressure of air

$$\Delta\alpha = 8.700806 - 8.598755 = 0.10205° = 6.12 MOA,$$

The change in departure angle that corresponds to an opposite range wind of 10m/s is

$$\Delta\alpha = 8.774658 - 8.598755 = 0.176° = 10.55 MOA.$$

There is no change in the departure angle because of the cross wind. The projectile deviates 84 meters in the direction of the z-axis, as result of the cross wind (10m/s).

The change in departure angle as result of a 10 degree Celsius increase in the temperature of powder load (black powder) is

$$\Delta\alpha = 8.338257 - 8.598755 = -0.2605° = -15.63 MOA.$$

The change in departure angle caused by the increase in projectile mass (input 27.30, and 0.180) is

$$\Delta\alpha = 8.605469 - 8.598755 = 0.0007° = 0.403 MOA.$$

The change in departure angle caused by the decrease in the initial speed of projectile (input initial speed 876.15) is

$$\Delta\alpha = 8.81628 - 8.598755 = 0.218° = 13.05 MOA.$$

Total Change in departure angle

$$\Delta\alpha_T = 2.653 MOA = 0.044217°.$$

The departure angle needed to hit the target as result of all changes is approximately

$$\alpha_0 = 8.598755 + 0.044217 = 8.642972°.$$

Example 2.25 Adjusted Rifleman's Rule

The Range table of the 122mm cannon, Mod. 1960 (Standard Atmosphere), shows that to the horizontal range 10000 meters corresponds the departure angle 6.2 degrees.

Use the Adjusted Rifleman's Rule to find the departure angle needed to hit the target located on a 15 degree inclined plane at a point with abscissa x = 10000 meters. Verify the result using Angmet.Bas, or Angle122.Bas.

Solution

(Ref. Section 1.6, Adjusted Rifleman's Rule, equations (1.6.2)-(1.6.4)) Substituting in equation (1.6.2),

$$\sin(2\overline{\alpha}_0 + E) = \sin(2\alpha_0)\cos(E) + \sin(E),$$

we obtain:

$$\sin(2\overline{\alpha}_0 + 15) = \sin(2 \cdot 6.20)\cos(15) + \sin(15) = 12.7903367$$

Hence, we find that the super elevation angle is

$$\overline{\alpha}_0 = (\sin^{-1}(12.7903367) - 15)/2 = 6.395168°.$$

The drop of the projectile fired at angle 6.2 degree is

$$\overline{y}_0 = 10000 \cdot \tan(6.2) = 1086.4m.$$

Using (1.6.3), for the drop of the projectile at the given inclined range, we have:

$$\overline{y}_T = \frac{\cos^2(\alpha_0)}{\cos^2(\overline{\alpha}_0 + E)}\overline{y}_0 = \frac{\cos^2(6.2)}{\cos^2(6.395168 + 15)} \cdot (1086.35) = 1238.5m,$$

Using Angmet.Bas for the point of impact (10000, 2679.5) at the inclined range, we have:

Input: x-coordinate, 10000; y-coordinate, 2679.5, speed, 885; and all the standard values, we find that the accurate departure angle is 21.54175 degree.

The accurate super elevation angle is 6.54175 degree, while the projectile drop is

$$\bar{y}_T = 10000[\tan(21.54175) - \tan(15)] = 1268m.$$

Remark

Comparing the results, we see that the adjusted rifleman's rule does not give accurate values for the super elevation angle and the projectile drop, when the horizontal range of fire is large.

The error in super elevation angle is -8.79 MOA, while the error in the projectile drop is (-30) meters. The discrepancies are result of the fact that the ballistics coefficient changes with the departure angle, and cannot be considered constant as it is supposed in Rifleman's rule.

QBasic PC Program ANGMET.BAS
122mm Russian Cannon, Mod. 1960
Non Standard Atmosphere, Wind Present

'Find: Departure Angle, Time of Flight, Impact Speed, Impact Angle, etc.
      'Given: The coordinates of the target and the location of the muzzle
      of the cannon
      '& Ballistics Coefficient as function of departure angle.
      'Temperature of air and thrusting charge of projectile are known;
      'The mass of projectile and the air humidity are known

'_____

'CONTROL DATA
'INPUT: x0 = 10000, projectile speed = 885
    'y-coordinate of cannon, 0, Temperature of air = 15, Temperature of
    propellant = 15
    'Pressure = 750, Pressure of Water vapor = 6.35, projectile mass = 27.3,
    change in mass, 0
    'Range wind = 10, cross wind, 10.
'

'RESULTS: Departure Angle = 6.12134 Degree, Time of Flight = 16.57
    'Impact Speed = 419, Impact Angle = -10108 Degree
    'Coordinates of the trajectory vertex (5611, 340), cross wind deflection, 52

'_____

'Functions and Sub. Prog.
```
DECLARE SUB y1z1v1w1 (x, y, z, v, w, y1, z1, v1, w1, koef, pa1, wind, ys,
    yy, pa, ta1)
DECLARE SUB InfHyres (xx, voo, vo1, ta, ta1, pa, pa1, ea, m, dm, tp, wind,
    xc, yc, vo, yc1, cw, xx1, koef)
DECLARE SUB InfDales (x0, y0, z0, v0, w0, xc, yc, A, aT, vo, yc1, xm, ym,
    cw, xx, voo, xa, ya, ta, va, aa)
DECLARE SUB NPxyzvw (nk, x, x0, y, y0, z, z0, v, v0, w, w0, h, h0, k, l, r, q)
DECLARE SUB NPkoef (k, l, r, q, h, y1, z1, v1, w1)
DECLARE SUB menu (cog, cof, xf, yf, xfu, yfu, t$)
DECLARE SUB y0z0 (y0, z0, A, vo1, wind)
DECLARE SUB c (koef, A, m, dm)
'Variables
SCREEN 0
1 :
DIM m(4, 4), v(4)          'Intermediate values (k,l,r,q)
rendi = 4                  'rend dif.
```

```
cog = 7: cof = 0
cikli = 0
A = 23                   'Initial Angle 23 degree
kendi = 22                'Angle [Degree] for maximum distance
kov = 1                  'Test of the value of v0
gab = 1                  'error 0.1 m.
tt = 1


'Solution
CLS
'Initial Data
menu cog, cof, 3, 10, 7, 70, "DATA INPUT"
InfHyres xx, voo, vol, ta, ta1, pa, pa1, ea, m, dm, tp, wind, xc, yc, yc1, vo, cw,
    xx1, koef
hap = 1
'Initial values
f:
x0 = 0: v0 = vo: w0 = 0
y0z0 y0, z0, A, vol, wind: h0 = hap
c koef, A, m, dm
ff:
FOR nk = 1 TO rendi
NPxyzvw nk, x, x0, y, y0, z, z0, v, v0, w, w0, h, h0, k, l, r, q
y1z1v1w1 x, y, z, v, w, y1, z1, v1, w1, koef, pa1, wind, ys, yy, pa, ta1
NPkoef k, l, r, q, h, y1, z1, v1, w1
m(nk, 1) = k: m(nk, 2) = l
m(nk, 3) = r: m(nk, 4) = q
NEXT nk


'Estimation for new points
FOR i = 1 TO rendi
v(i) = 1 / 6 * (m(1, i) + 2 * m(2, i) + 2 * m(3, i) + m(4, i))
NEXT i


'New Points
x0 = x0 + h: y0 = y0 + v(1): z0 = z0 + v(2)
v0 = v0 + v(3): w0 = w0 + v(4)

xmm = x0 + wind * w0
IF ABS(z0) <= .0001 THEN
```

```
xm = xmm
ym = v0
END IF


xaa = x0 + wind * w0
IF ABS(xx1 - xaa) <= 1 THEN
xa = xaa
ya = v0
ta = w0
va = y0 / COS(ATN(z0))
aa = (180 / 3.1415169542#) * ATN(z0)
END IF

'Tests the y-value
xT = x0 + wind * w0
IF kov = 1 THEN kov = -1: GOTO ff:
IF ABS(xT - xc) < gab AND ABS(v0 - yc) <= (gab * TAN(A * 3.1415954# / 180))
    THEN
c:
'DISPLAY of RESULTS
CLS
PLAY "a8a16a32b8"
menu 12, 0, 5, 10, 11, 70, "RESULTS:"
COLOR 7
InfDales x0, y0, z0, v0, w0, A, xc, yc, aT, vo, yc1, xm, ym, cw, xx, voo, xa,
    ya, ta, va, aa
CLS
GOTO 1:
END IF
IF ABS(xT - xc) < gab AND (v0 - yc) > (gab * TAN(A * 3.1415954# / 180)) THEN
t$ = "* ? *"
menu 18, 0, 10, 20, 14, 60, t$
COLOR 14
LOCATE 12, 30: PRINT "Wait a moment, Please (+)";
LOCATE 12, 53: PRINT tt
tt = tt + 1
COLOR 7
A = A - kendi
```

```
GOTO fff:
END IF

IF ABS(xT - xc) < gab AND (v0 - yc) < ((-1 * gab) * TAN(A * 3.1415954# / 180))
    THEN
t$ = "* ? *"
menu 18, 0, 10, 20, 14, 60, t$
COLOR 14
LOCATE 12, 30: PRINT "Wait a moment, Please (-)";
LOCATE 12, 53: PRINT tt
tt = tt + 1
COLOR 7
A = A + kendi
GOTO fff:
END IF
GOTO ff:

fff:
'Restart Cycle
cikli = cikli + 1
IF cikli = 20 THEN GOTO c:
kendi = kendi / 2
kov = 1
GOTO f:

SUB c (koef, A, m, dm)
IF A >= 0 AND A <= 1.38333333# THEN koef = (1 - dm / m) * (1.82051 - 210.432
    * (A * 3.141592654# / 180) + 10066.5 * (A * 3.141592654# / 180) ^
    2 - 164950 * (A * 3.141592654# / 180) ^ 3) '3200
IF A > 1.38333333# AND A <= 2.5666667# THEN koef = (1 - dm / m)
    * (.490432 - 14.0538 * (A * 3.141592654# / 180) + 298.238 * (A *
    3.141592654# / 180) ^ 2 - 2328.38 * (A * 3.141592654# / 180) ^ 3) '5400
IF A > 2.5666667# AND A <= 4.8833333# THEN koef = (1 - dm / m)
    * (.352739 - 4.0664 * (A * 3.141592654# / 180) + 49.8947 * (A *
    3.141592654# / 180) ^ 2 - 207.518 * (A * 3.141592654# / 180) ^ 3) '8600
IF A > 4.8833333# AND A < 8.0066667# THEN koef = (1 - dm / m) *
    (.304165 - 1.65929 * (A * 3.141592654# / 180) + 13.5957 * (A *
    3.141592654# / 180) ^ 2 - 35.3659 * (A * 3.141592654# / 180) ^ 3) '11600
```

```
IF A > 8.0066667# AND A <= 12.9166667# THEN koef = (1 - dm / m)
    * (.324563 - 1.5842 * (A * 3.141592654# / 180) + 9.56496 * (A *
    3.141592654# / 180) ^ 2 - 17.802 * (A * 3.141592654# / 180) ^ 3)'14600
IF A > 12.9166667# AND A <= 18.3166667# THEN koef = (1 - dm / m)
    * (.158193 + .919854 * (A * 3.141592654# / 180) - 3.03889 * (A *
    3.141592654# / 180) ^ 2 + 3.35924 * (A * 3.141592654# / 180) ^ 3) '17200
IF A > 18.3166667# AND A <= 45 THEN koef = (1 - dm / m) *
    (.251064 - .0064118# * (A * 3.141592654# / 180) + .0262451# * (A *
    3.141592654# / 180) ^ 2 - .0064443# * (A * 3.141592654# / 180) ^ 3) '23800
END SUB


SUB InfDales (x0, y0, z0, v0, w0, A, xc, aT, yc, vo, yc1, xm, ym, cw, xx, voo,
    xa, ya, ta, va, aa)
aT = ATN(z0) * 180 / 3.141592654#
LOCATE 6, 16: PRINT "Departure Angle        :"; A; "Degree"
LOCATE 7, 16: PRINT "Time of Flight to Target   :"; INT((w0) * 1000 + .5)
    / 1000
LOCATE 8, 16: PRINT "Terminal Speed         :"; INT((y0 / COS(ATN(z0)))
    * 100 + .5) / 100; "m/s"
LOCATE 9, 16: PRINT "Terminal Angle         :"; aT; "degree"
LOCATE 10, 16: PRINT "Coordinates of Vertex    :"; "("; INT((xm) * 100 +
    .5) / 100; ",", "; INT((ym) * 100 + .5) / 100; ")"
LOCATE 11, 16: PRINT "Cross-Wind Deflection    :"; INT((cw * (w0 - xx /
    (voo * COS(A * 3.14159265# / 180)))) * 1000 + .5) / 1000
LOCATE 13, 16: PRINT "Location of TARGET       :"; "("; xx; yc1; ")"
LOCATE 14, 16: PRINT "Location of FIREARM      :"; "(0,"; vo; ")"
LOCATE 15, 16: PRINT "Distance to TARGET       :"; (xx ^ 2 + (yc1 - vo) ^ 2) ^ .5
LOCATE 17, 16: PRINT "X-Coordinate of the Point  :"; INT((xa) * 100 + .5) / 100
LOCATE 18, 16: PRINT "Y-Coordinate of the Point  :"; INT((ya) * 100 + .5) / 100
LOCATE 19, 16: PRINT "Time of Flight to the Point :"; INT((ta) * 100 + .5) / 100
LOCATE 20, 16: PRINT "Speed at the Point       :"; INT((va) * 100 + .5) / 100
LOCATE 21, 16: PRINT "Angle at the Point       :"; aa
LOCATE 22, 16: PRINT "Cross Wind Deflection      :"; INT((cw * (ta - xa /
    (voo * COS(A * 3.14159265# / 180)))) * 1000 + .5) / 1000
COLOR 7
LOCATE 24, 11: PRINT "Pres [ P ] to repeat [ Esc ] to end";
cc$ = INPUT$(1)
IF cc$ = CHR$(27) THEN SCREEN 9: CLS : END
END SUB
```

```
SUB InfHyres (xx, voo, vo1, ta, ta1, pa, pa1, ea, m, dm, tp, wind, xc, yc, yc1,
    vo, cw, xx1, koef)
LOCATE 4, 13: INPUT "x-coordinate of TARGET [m]     = "; xx
LOCATE 5, 13: INPUT "y-Coordinate of TARGET [m]     = "; yc

LOCATE 6, 13: INPUT "y-Coordinate of FIREARM [m]     = "; vo
LOCATE 7, 13: INPUT "Departure Speed [m/s]       = "; voo
LOCATE 9, 13: INPUT "Temperature of Air [C]       = "; ta
LOCATE 10, 13: INPUT "Propellant Temperature [C]   = "; tp
LOCATE 11, 13: INPUT "Atmospheric Pressure [mm Hg] = "; pa
LOCATE 12, 13: INPUT "Pressure of Water vapor [mm Hg]= "; ea
LOCATE 13, 13: INPUT "Projectile Standard Mass [kg] = "; m
LOCATE 14, 13: INPUT "Change in Projectile Mass    = "; dm
LOCATE 15, 13: INPUT "Range-Wind [m/s]           = "; wind
LOCATE 16, 13: INPUT "Cross-Wind [m/s]          = "; cw
LOCATE 17, 13: INPUT "X-Coordinate of a Point [m]  = "; xx1
CLS
xc = xx
yc1 = yc

ta = ta + 273.15
pa1 = ta / (1 - .3785 * ea / pa)
vo1 = (voo - .4 * voo * (dm / m) + .00125 * voo * (tp - 15))
END SUB


SUB menu (cog, cof, xf, yf, xfu, yfu, t$)
COLOR cog, cof
LOCATE xf - 1, yf: PRINT t$
LOCATE xf, yf: PRINT "É" + STRING$(yfu - yf, 205) + "»";
FOR i = xf + 1 TO xfu
LOCATE i, yf: PRINT "º" + SPACE$(yfu - yf) + "º";
NEXT
LOCATE xfu + 1, yf: PRINT "È" + STRING$(yfu - yf, 205) + "¼";
END SUB


SUB NPkoef (k, l, r, q, h, y1, z1, v1, w1)
k = h * y1: l = h * z1
r = h * v1: q = h * w1
END SUB
```

```
SUB NPxyzvw (nk, x, x0, y, y0, z, z0, v, v0, w, w0, h, h0, k, l, r, q)
IF nk = 1 THEN
x = x0: y = y0: z = z0
v = v0: w = w0: h = h0
GOTO fund:
END IF


IF nk = 2 OR nk = 3 THEN
x = x0 + (.5 * h): y = y0 + (.5 * k)
z = z0 + (.5 * l): v = v0 + (.5 * r)
w = w0 + (.5 * q)
GOTO fund:
END IF


IF nk = 4 THEN
x = x0 + h: y = y0 + k: z = z0 + l
v = v0 + r: w = w0 + q
END IF
fund:
END SUB


SUB y0z0 (y0, z0, A, vo1, wind)
y0 = SQR(vo1 ^ 2 + wind ^ 2 - 2 * vo1 * wind * COS(A * 3.141592654# / 180))
y0 = y0 * COS(A * 3.141592654# / 180)
z0 = TAN(A * 3.141592654# / 180)
z0 = z0 / (1 - wind / (vo1 * COS(A * 3.141592654# / 180)))
END SUB


SUB y1z1v1w1 (x, y, z, v, w, y1, z1, v1, w1, koef, pa1, wind, ys, yy, pa, ta1)
yy = y * SQR(1 + z ^ 2)
ta1 = (289.08 / pa1) ^ .5
IF (yy * ta1) >= 256 THEN
y1 = -1 * koef * (pa / 750) * (((pa1 - .006328 * v) / pa1) ^ 4.4) * (ta1 *
    yy - 240) / (3 * yy)
ELSE
y1 = -1 * 1.212 * 10 ^ - 4 * koef * (pa / 750) * (((pa1 - .006328 * v) / pa1) ^
    4.4) * (ta1 * yy) ^ 2 / (y * SQR(1 + z ^ 2))
END IF
z1 = -9.80665 / y ^ 2
v1 = z
w1 = 1 / y
END SUB
```

2.6 PC Program ANATO762.Bas, 7.62mm NATO Bullet

The PC program ANATO762.Bas can be used to find the departure angle of the caliber 7.622mm M852 HPBT NATO bullet to hit a target at a given location. We assume a standard atmosphere, and a bullet that have the nominal characteristics, i.e. the mass 0.0109kg (168grain), caliber 7.62mm, initial speed 807.72m/s, (2650fps).

The ballistics coefficient is given as a function of the projectile speed. We get approximately the same results if we employ the AngleC.Bas and a ballistics coefficient equal to 2.945 (see section 1.1).

Use of the ANATO762.Bas

Example 2.26 Adjusted Rifleman's Rule

A bullet caliber 7.62mm M852 HPBT is fired from a Winchester rifle with an initial speed 807.70m/s. Use the PC Program ANATO762.Bas.

(a) Find the departure angle needed to zero the rifle at the horizontal range 500 meters, and find the drop of the bullet respectively at horizontal ranges 300m and 500m.
(b) Find the super elevation angle if the target is located at the point with coordinates (500m, 200m) while the rifle is at the sea level.
(c) Find the super elevation angle if the site angle is 30 degree, and the target is at the inclined range 500 meters.

Solution

(a) **Input**: x-coordinate of target, 500; departure speed, 807.70; abscissa x of a trajectory point, 300.
    Output: Departure Angle, 0.299072 degree (17.94 MOA); Time, 0.787s; speed, 502s; terminal angle, -0.4098; vertex, (270m, 0.76m)

For the horizontal range 300m, we find:
y-coordinate, 0.749m; speed, 614m/s; terminal angle, -0.0437 degree. The projectile drop at 500 meters is

$$\bar{y}_s = x \tan \alpha_0 = 500 \cdot \tan(0.299072) = 2.61m \,.$$

The projectile drop at 300 meters is

$$\overline{y}_3 = x_3 \tan\alpha_0 - y_3 = 300 \cdot \tan(0.299072) - 0.749 = 0.817m \, .$$

(b) Input x-coordinate of target, 500; y-coordinate of target 200; departure speed 807.72. We obtain the departure angle 22.10907degree; the inclined distance to the target, 538.50m. The site angle is

$$E = \tan^{-1}(200/500) = 21.80141° \, .$$

The super elevation angle is

$$\overline{\alpha}_0 = 22.10907 - 21.80141 = 0.30766° = 18.45 MOA \, .$$

Remark

We see that the super elevation angle (18.45MOA) that corresponds to the abscissa 500 meters is slightly different from the horizontal departure angle (17.94 MOA) that corresponds to the horizontal range 500 meters.
It can be easily verified that the results are compatible with the Adjusted Rifleman's Rule (see formula (1.6.2), section 1.6):

$$\sin(2\overline{\alpha}_0 + E) = \sin(2\alpha_0)\cos(E) + \sin(E) \, .$$

(c) The coordinates of the target are (433, 250). Using the PC program ANATO762.Bas, we find that the departure angle is 30.21875 degree. The super elevation angle is 0.21875 degree (13.31 MOA).

Remark

It can be seen that the super elevation angle (15.70 MOA) that corresponds to the inclined range 500 meters can be obtained from the departure angle (17.92 MOA) that corresponds to the horizontal range 500 meters using the formula (1.7.5):

$$\overline{\alpha}_0 \approx \alpha_0 \cdot \cos(E) \, .$$

PC Program ANATO762.Bas
Standard Atmosphere, Wind is Present
'NATO Rifle (0.308 Winchester), Bullet 7.62mm M852 HPBT,
'Projectile Initial speed, 2650 ft/s (807.72m/s, 168 grain, BC = 2.945
'Given: Coordinates of the Target and the firearm,
'Ballistics Coefficient—Function of angle
'Find: Launching Angle, Time of Flight, Impact Speed, Impact Angle, etc
'

'CONTROL DATA
'INPUT: x-coordinate of target = 300 m, projectile initial speed = 807.72
'RESULTS: Departure Angle = 0.1555214, Degree, Time of
            Flight = 0.426s
            'Impact Speed = 614m/s, Impact Angle = -0.18686 Degree
            'Vertex (157, 0.22)
'

'Note: Round the Input RANGE to the nearest 1; neglect the numbers after decimal
'

'Functions and Sub. Prog.

DECLARE SUB y1z1v1w1 (x, y, z, v, w, y1, z1, v1, w1, koef, ys, yy, wind)
DECLARE SUB InfHyres (xx, n, koef, vv, vo, yo, xc1, cw, wind)
DECLARE SUB InfDales (x0, y0, z0, v0, w0, xm, ym, A, xc, yc, tc, vc, ac, vo, vv, xx, vo, cw)
DECLARE SUB NPxyzvw (nk, x, x0, y, y0, z, z0, v, v0, w, w0, h, h0, k, l, r, q)
DECLARE SUB NPkoef (k, l, r, q, h, y1, z1, v1, w1)
DECLARE SUB menu (cog, cof, xf, yf, xfu, yfu, t$)
DECLARE SUB y0z0 (y0, z0, A, vo, wind)
DECLARE SUB c (koef, y0, A)

'Variables
SCREEN 0
1 :
DIM m(4, 4), v(4)   'Intermediate values (k,l,r,q)
rendi = 4   'rend dif.
cog = 7: cof = 0
cikli = 0
A = 23   'Initial Angle 23 degree
kendi = 22   'd.Angle for maximum distance
kov = 1   'Test of the value of v0
tt = 1

```
'Solution
CLS
'Initial Data
menu cog, cof, 3, 10, 7, 70, "DATA INPUT"
InfHyres xx, n, koef, vv, vo, yo, xc1, cw, wind
hap = 1
IF wind = 0 THEN
gab = .01
ELSE
gab = 1
END IF

'Initial values
f:
x0 = 0: v0 = vv: w0 = 0
y0z0 y0, z0, A, vo, wind: h0 = hap
c koef, y0, A
ff:
FOR nk = 1 TO rendi
NPxyzvw nk, x, x0, y, y0, z, z0, v, v0, w, w0, h, h0, k, l, r, q
y1z1v1w1 x, y, z, v, w, y1, z1, v1, w1, koef, ys, yy, wind
NPkoef k, l, r, q, h, y1, z1, v1, w1
m(nk, 1) = k: m(nk, 2) = l
m(nk, 3) = r: m(nk, 4) = q
NEXT nk

'Estimations for new points
FOR i = 1 TO rendi
v(i) = 1 / 6 * (m(1, i) + 2 * m(2, i) + 2 * m(3, i) + m(4, i))
NEXT i

'New Points
x0 = x0 + h: y0 = y0 + v(1): z0 = z0 + v(2)
v0 = v0 + v(3): w0 = w0 + v(4)

xcc = x0 + wind * w0
IF ABS(xcc - xc1) <= .5 THEN

xc = xcc
yc = v0
```

```
tc = w0
ac = (180 / 3.141592654#) * ATN(z0)
vc = y0 / COS(ATN(z0))
END IF


xmm = x0 + w0 * wind
IF xmm > 30 AND ABS(z0) <= .00001 THEN
xm = xmm
ym = v0
END IF


'Tests the y-value
xT = x0 + wind * w0
IF kov = 1 THEN kov = -1: GOTO ff:
IF ABS(xT - xx) < gab AND v0 <= yo + (gab * TAN(A * 3.1415954# / 180))
    AND v0 >= yo + ((-1 * gab) * TAN(A * 3.1415954# / 180)) THEN
c:
'Display Results
CLS
PLAY "a8a16a32b8"
menu 12, 0, 5, 10, 11, 70, "RESULTS:"
COLOR 7
InfDales x0, y0, z0, v0, w0, xm, ym, A, xc, yc, tc, vc, ac, yo, vv, xx, vo, cw
CLS
GOTO 1:
END IF


IF ABS(xT - xx) < gab AND v0 > yo + (gab * TAN(A * 3.1415954# / 180)) THEN
t$ = "* ? *"
menu 18, 0, 10, 20, 14, 60, t$
COLOR 14
LOCATE 12, 30: PRINT "Wait a moment, Please (+)";
LOCATE 12, 53: PRINT tt
tt = tt + 1
COLOR 7
A = A - kendi
GOTO fff:
END IF


IF ABS(xT - xx) < gab AND v0 < yo + ((-1 * gab) * TAN(A * 3.1415954# / 180))
    THEN
```

```
t$ = "* ? *"
menu 18, 0, 10, 20, 14, 60, t$
COLOR 14
LOCATE 12, 30: PRINT "Wait a moment, Please (-)";
LOCATE 12, 53: PRINT tt
tt = tt + 1
COLOR 7
A = A + kendi
GOTO fff:
END IF
GOTO ff:

fff:
'Restart Cycle
cikli = cikli + 1
IF cikli = 20 THEN GOTO c:
kendi = kendi / 2
kov = 1
GOTO f:

SUB c (koef, y0, A)
koef = 1.0563 * (1.2112105619# + .01289224624# * y0 - .0000261226# * y0
   ^ 2 + .000000015573# * y0 ^ 3)
END SUB

SUB InfDales (x0, y0, z0, v0, w0, xm, ym, A, xc, yc, tc, vc, ac, yo, vv, xx, vo, cw)
aT = ATN(z0) * 180 / 3.141592654#
LOCATE 6, 16: PRINT "Launching Angle    = "; A; "degree"
LOCATE 7, 16: PRINT "Time of Flight  = "; INT((w0) * 1000 + .5) / 1000; "seconds";
LOCATE 8, 16: PRINT "Terminal Speed     = "; INT((y0 / COS(ATN(z0))) *
   100 + .5) / 100; "m/s"
LOCATE 9, 16: PRINT "Terminal Angle     = "; aT; "degree"
LOCATE 10, 16: PRINT "Trajectory Vertex  = :"; "("; INT((xm) * 100 + .5) /
   100; ","; INT((ym) * 100 + .5) / 100; ")"
LOCATE 11, 16: PRINT "Cross-Wind Deflection     = "; INT((cw * (w0 - xx
   / (vo * COS(A * 3.14159265# / 180))))) * 1000 + .5) / 1000
LOCATE 13, 16: PRINT "Distance to the target    = "; INT(((xx ^ 2 + (vv - v0)
   ^ 2) ^ .5) * 100 + .05) / 100; "meter";
LOCATE 14, 16: PRINT "x, y coordinates of TARGET = "; "("; INT((xx) *
   100 + .5) / 100; ","; INT((v0) * 100 + .5) / 100; ")"
```

```
LOCATE 15, 16: PRINT "x,y coordinatees of GUN    = "; "("; 0; ","; vv; ")"
LOCATE 17, 18: PRINT "x-coordinate of a point[m] :"; INT((xc) * 100 + .5) / 100
LOCATE 18, 18: PRINT "Corresponding y [m]     :"; INT((yc) * 1000 + .5) / 1000
LOCATE 19, 18: PRINT "Corresponding Time [sec]  :"; INT((tc) * 100 + .5) / 100
LOCATE 20, 18: PRINT "Corresponding Speed [m/s] :"; INT((vc) * 100 + .5) / 100
LOCATE 21, 18: PRINT "Corresponding Angle [Deg]   :"; ac
LOCATE 22, 18: PRINT "Cross-Wind Deflection      :"; INT((cw * (tc - xc /
    (vo * COS(A * 3.14159265# / 180)))) * 1000 + .5) / 1000
COLOR 7
LOCATE 24, 11: PRINT "Pres [ P ] to repeat [ Esc ] to end  ";
cc$ = INPUT$(1)
IF cc$ = CHR$(27) THEN SCREEN 9: CLS : END
END SUB


SUB InfHyres (xx, n, koef, vv, vo, yo, xc1, cw, wind)
LOCATE 4, 13: INPUT "x-coordinate of TARGET [m] ="; xx
LOCATE 5, 13: INPUT "y-coordinate of TARGET [m] ="; yo
LOCATE 6, 13: INPUT "y-coord of GUN ="; vv
LOCATE 7, 13: INPUT "Initial Speed [m/s] ="; vo
LOCATE 9, 13: INPUT "X-coordinate of a point ="; xc1
LOCATE 10, 13: INPUT "Range-Wind ="; wind
LOCATE 11, 13: INPUT "Cross-Wind ="; cw
CLS
END SUB


SUB menu (cog, cof, xf, yf, xfu, yfu, t$)
COLOR cog, cof

LOCATE xf - 1, yf: PRINT t$
LOCATE xf, yf: PRINT "É" + STRING$(yfu - yf, 205) + "»";
FOR i = xf + 1 TO xfu
LOCATE i, yf: PRINT """ + SPACE$(yfu - yf) + """;
NEXT
LOCATE xfu + 1, yf: PRINT "È" + STRING$(yfu - yf, 205) + "¼";
END SUB


SUB NPkoef (k, l, r, q, h, y1, z1, v1, w1)
k = h * y1: l = h * z1
r = h * v1: q = h * w1
END SUB
```

```
SUB NPxyzvw (nk, x, x0, y, y0, z, z0, v, v0, w, w0, h, h0, k, l, r, q)
IF nk = 1 THEN
x = x0: y = y0: z = z0
v = v0: w = w0: h = h0
GOTO fund:
END IF

IF nk = 2 OR nk = 3 THEN
x = x0 + (.5 * h): y = y0 + (.5 * k)
z = z0 + (.5 * l): v = v0 + (.5 * r)
w = w0 + (.5 * q)
GOTO fund:
END IF

IF nk = 4 THEN
x = x0 + h: y = y0 + k: z = z0 + l
v = v0 + r: w = w0 + q
END IF
fund:
END SUB


SUB y0z0 (y0, z0, A, vo, wind)
y0 = SQR(vo ^ 2 + wind ^ 2 - 2 * vo * wind * COS(A * 3.141592654# / 180))
y0 = y0 * COS(A * 3.141592654# / 180)
z0 = TAN(A * 3.141592654# / 180)
z0 = z0 / (1 - wind / (vo * COS(A * 3.141592654# / 180)))
END SUB


SUB y1z1v1w1 (x, y, z, v, w, y1, z1, v1, w1, koef, ys, yy, wind)
yy = y * SQR(1 + z ^ 2)
IF yy > 256! THEN
y1 = -1 * koef * ((289.08 - .006328 * v) / 289.08) ^ 4.4 * (yy - 240) / (3 * yy)
ELSE
y1 = -1 * koef * ((289.08 - .006328 * v) / 289.08) ^ 4.4 * .0001212 * yy ^ 2 / yy
END IF
z1 = -9.80665 / y ^ 2
v1 = z
w1 = 1 / y
END SUB
```

2.7 PC Program Ingaan.Bas, English Units

The PC program **Ingaan.Bas** is similar to **Anglec.Bas** and is prepared to be employed by ballisticians that are used and prefer to make the calculations in English units (Ref. G. Klimi, "Exterior Ballistics with Applications", p.79, Xlibris, 2008). The program uses the approximate Siacci function that replaces the Ingall's drag function. The Siacci ballistics coefficient C in English units that corresponds to the referred drag function is calculated using formula 2.4.9 of the above reference book,

$$C = \frac{1.4223}{c} , \qquad\qquad (2.1)$$

where c is the ballistics coefficient in SI units.

The PC program **Ingaan.Bas** can be used to determine the departure angle of a projectile needed to hit a given target when are known:

The ballistics coefficient, projectile departure speed, range of fire.

The projectile is launched in the Standard Atmosphere and **the target and the gun are at the sea level**.

The program estimates the coordinates of the trajectory vertex, as well as the time of flight to the target, the terminal speed and the terminal angle. The program finds as well the trajectory elements of a point on the trajectory for a given x-coordinate of that point. The value of the x-coordinate of the given point must be less than the projectile range.

The program uses a constant BC that represents an average ballistics coefficient or one that usually is given in ballistics tables, or in the handbook of the firearm.

- For some bullets, listed in section 1.1, we can find the Ingalls ballistics coefficients using (2.1). For example, the Ingalls ballistics coefficient of a 7.62mm M852 HPBT bullet (departure speed 2600, with a Siacci BC = 2.945) is

$$C = \frac{1.4223}{c} = \frac{1.4223}{2.945} = 0.483 .$$

Note: The program Ingaan.Bas can be easily modified to consider the sight height of a rifle.

Example 2.27

The average ballistics coefficient of the caliber 0.30 Ball M2 bullet, estimated in Exercise 2, page 98, of the book "Exterior Ballistics with Applications, by G. Klimi, Xlibris, 2008 is 3.384773 (it corresponds to the average form coefficient 0.538).

The ballistics coefficient in English units that can be calculated easily employing formula (2.1) is $C = 0.4202 lb / in$.

Use the above average ballistics coefficient $C = 0.4202 lb / in$ to find the departure angle needed to hit a target located at the range:

(a) 600 feet; (b) 1200 feet; (c) 1800 feet

The departure speed of the projectile is 2800fps. The projectile flight is in standard atmosphere at the sea level.

Solution

(a) Input: Range, 200 yard; projectile speed, 2800; ballistics coefficient 0.4202.

  Note: We may input as well the x-coordinate of a point on the projectile trajectory, for example 100 yard.

  Output: For the range 200 ft.:
  Angle is 0.07986 degree; time is 0.235s; speed is 2331fps; impact angle,-0.09; trajectory vertex is located at (329ft, 0.22ft).

In a similar way, we find:

(b) For the range 400yd:
  Angle is 0.18292 degree; time is 0.52s; speed is 1907fps; impact angle,-0.23674; trajectory vertex is located at (654ft, 1.08ft).

(c) For the range 600ft.:
  Angle is 0.3206 degree; time is 0.87s; speed is 1541fps; impact angle,-0.47793; trajectory vertex is located at (1002ft, 3.06ft).

PC QuickBasic Program INGAAN.BAS
Standard Atmosphere, Wind is not Present
'English Units
'FIND: DEPARTURE Angle, Time of Flight, y-coordinate of a point
'GIVEN: RANGE, Ballistics Coefficient "C=m/id^2", Departure Speed
  (Standard Atmosphere)
'FIND: LAUNCHING Angle, Time of Flight, y-coordinate of a point
'GIVEN: RANGE, Ballistics Coefficient "C=m/id^2", Launching Speed
  (Standard Atmosphere)
'_____

CONTROL DATA
'INPUT: Range = 500yd: Departure Speed = 2,600: Ballistics Coefficient=0.483,
         'x-coordinate of a point on Trajectory = 300
'RESULTS: Departure Angle = .27557 [Degree], Time of Flight = 0.719 [s]
         'Terminal Angle = -0.36902 [Degree], Terminal speed = 1682[ft./s]
         'Coordinates of Trajectory vertex [ft] (817, 2.08)

         'For x =200yd: y = 1.93ft., Time = 0.25s,
                  'Angle = 0.08178 Degree, Speed = 2204ft/s.
'_____

'FUNCTIONS, SUBS.
DECLARE SUB y1z1v1w1 (x, y, z, v, w, y1, z1, v1, w1, koef)
DECLARE SUB InfHyres (xx, koef, Speed, xc1)
DECLARE SUB InfDales (x0, y0, z0, v0, w0, xc, yc, tc, ac, vc, xm, ym, a)
DECLARE SUB NPxyzvw (nk, x, x0, y, y0, z, z0, v, v0, w, w0, h, h0, k, l, r, q)
DECLARE SUB NPkoef (k, l, r, q, h, y1, z1, v1, w1)
DECLARE SUB menu (cog, cof, xf, yf, xfu, yfu, t$)
DECLARE SUB y0z0 (y0, z0, Speed, a)
DECLARE SUB c (koef, a)
'Variables
SCREEN 0
1 :
DIM m(4, 4), v(4)
rendi = 4
cog = 7: cof = 0
cikli = 0
a = 23   'Initial Guessed Angle 23
kendi = 22
kov = 1

```
gab = 1.5
tt = 1

'Solution
CLS
'Initial Data
menu cog, cof, 3, 10, 8, 70, "INPUT"
InfHyres xx, koef, Speed, xc1
hap = 1

'Start
f:
x0 = 0: v0 = 0: w0 = 0
y0z0 y0, z0, Speed, a: h0 = hap
c koef, a
ff:
FOR nk = 1 TO rendi
NPxyzvw nk, x, x0, y, y0, z, z0, v, v0, w, w0, h, h0, k, l, r, q
y1z1v1w1 x, y, z, v, w, y1, z1, v1, w1, koef
NPkoef k, l, r, q, h, y1, z1, v1, w1
m(nk, 1) = k: m(nk, 2) = l
m(nk, 3) = r: m(nk, 4) = q
NEXT nk

'Calculation
FOR i = 1 TO rendi
v(i) = 1 / 6 * (m(1, i) + 2 * m(2, i) + 2 * m(3, i) + m(4, i))
NEXT i

'New Point
x0 = x0 + h: y0 = y0 + v(1): z0 = z0 + v(2)
v0 = v0 + v(3): w0 = w0 + v(4)

IF ABS(x0 - xc1) <= 1 THEN
xc = x0
yc = v0
tc = w0
ac = (180 / 3.141592654#) * ATN(z0)
vc = y0 / COS(ATN(z0))
```

```
END IF

IF ABS(z0) <= .0001 THEN
xm = x0
ym = v0
END IF

'Test if the y-coordinate passes here
IF kov = 1 THEN kov = -1: GOTO ff:
IF ABS(x0 - xx) < .01 AND v0 <= (gab * TAN(a * 3.1415954# / 180)) AND
    v0 >= ((-1 * gab) * TAN(a * 3.1415954# / 180)) THEN
c:
'Display of results
CLS
PLAY "a8a16a32b8"
menu 12, 0, 5, 10, 11, 70, "RESULTS:"
COLOR 7
InfDales x0, y0, z0, v0, w0, xc, yc, tc, ac, vc, xm, ym, a
CLS
GOTO 1:
END IF

IF ABS(x0 - xx) < .01 AND v0 > (gab * TAN(a * 3.1415954# / 180)) THEN
t$ = "* ? *"
menu 18, 0, 10, 20, 14, 60, t$
COLOR 14
LOCATE 12, 30: PRINT "Wait a Moment Please (+)";
LOCATE 12, 53: PRINT tt
tt = tt + 1
COLOR 7
a = a - kendi
GOTO fff:
END IF

IF ABS(x0 - xx) < .01 AND v0 < ((-1 * gab) * TAN(a * 3.1415954# / 180)) THEN
t$ = "* ? *"
menu 18, 0, 10, 20, 14, 60, t$
COLOR 14
LOCATE 12, 30: PRINT "Wait a Moment Please (-)";
```

```
LOCATE 12, 53: PRINT tt
tt = tt + 1
COLOR 7
a = a + kendi
GOTO fff:
END IF
GOTO ff:

fff:
'Restart Cycle
cikli = cikli + 1
IF cikli = 40 THEN GOTO c:
kendi = kendi / 2
kov = 1
GOTO f:
SUB c (koef, a)
koef = koef
END SUB

SUB InfDales (x0, y0, z0, v0, w0, xc, yc, tc, ac, vc, xm, ym, a)
g = (a - INT(a)) * .6 + INT(a)
zg = y0 / COS(ATN(z0))
z0 = (180 / 3.1415169542#) * ATN(z0)
sh = a / .06
LOCATE 6, 16: PRINT "Horizontal RANGE :"; x0 / 3; "yard"
LOCATE 7, 16: PRINT "LAUNCHING ANGLE :"; a; "Degree"
LOCATE 8, 16: PRINT "TIME OF FLIGHT :"; w0; "seconds"
LOCATE 9, 16: PRINT "TERMINAL SPEED :"; zg; "ft/s"
LOCATE 10, 16: PRINT "TERMINAL ANGLE :"; z0; "Degree"
LOCATE 12, 14: PRINT "Coordinates of Vertex (x, y) [ft] :"; "("; xm;", "; ym;")"
LOCATE 14, 18: PRINT "x-coordinate of a point [yd] :"; xc / 3
LOCATE 15, 18: PRINT "Corresponding y [ft] :"; yc
LOCATE 16, 18: PRINT "Corresponding Time [s] :"; tc
LOCATE 17, 18: PRINT "Corresponding Speed [ft /s] :"; vc
LOCATE 18, 18: PRINT "Corresponding Angle [Deg] :"; ac
COLOR 7
LOCATE 22, 15: PRINT "Press [ P ] to Repeat and [ Esc ] to End";
COLOR 10
```

```
LOCATE 22, 25: PRINT "P";
LOCATE 22, 45: PRINT "Esc";
COLOR 7
cc$ = INPUT$(1)
IF cc$ = CHR$(27) THEN SCREEN 9: CLS : END
END SUB


SUB InfHyres (xx, koef, Speed, xc1)
LOCATE 5, 13: INPUT "Horizontal RANGE [yard] ="; xx
xx = xx * 3
LOCATE 6, 13: INPUT "Initial Speed [ft/s] ="; Speed
LOCATE 7, 13: INPUT "Ballistics Coefficient ="; koef
koef = 1.4223 / koef
LOCATE 12, 13: INPUT "x-Coordinate of a point [yd] ="; xc1
xc1 = xc1 * 3
CLS
END SUB


SUB menu (cog, cof, xf, yf, xfu, yfu, t$)
COLOR cog, cof
LOCATE xf - 1, yf: PRINT t$
LOCATE xf, yf: PRINT "É" + STRING$(yfu - yf, 205) + "»";
FOR i = xf + 1 TO xfu
LOCATE i, yf: PRINT "°" + SPACE$(yfu - yf) + "°";
NEXT
LOCATE xfu + 1, yf: PRINT "È" + STRING$(yfu - yf, 205) + "¼";
END SUB


SUB NPkoef (k, l, r, q, h, y1, z1, v1, w1)
k = h * y1: l = h * z1
r = h * v1: q = h * w1
END SUB


SUB NPxyzvw (nk, x, x0, y, y0, z, z0, v, v0, w, w0, h, h0, k, l, r, q)
IF nk = 1 THEN
x = x0: y = y0: z = z0
v = v0: w = w0: h = h0
GOTO fund:
```

```
END IF

IF nk = 2 OR nk = 3 THEN
x = x0 + (.5 * h): y = y0 + (.5 * k)
z = z0 + (.5 * l): v = v0 + (.5 * r)
w = w0 + (.5 * q)
GOTO fund:
END IF

IF nk = 4 THEN
x = x0 + h: y = y0 + k: z = z0 + 1
v = v0 + r: w = w0 + q
END IF
fund:
END SUB

SUB y0z0 (y0, z0, Speed, a)
y0 = Speed * (COS(a * 3.141516954# / 180))
z0 = TAN(a * 3.141516954# / 180)
END SUB

SUB y1z1v1w1 (x, y, z, v, w, y1, z1, v1, w1, koef)
IF (y * SQR(1 + z ^ 2)) > 840! THEN
y1 = -1 * koef * EXP(-.0000315914# * v) * (1 / 3 - 262.1338667# / (y *
    SQR((1 + z ^ 2))))
ELSE
y1 = -1 * koef * EXP(-.0000315914# * v) * .0001212 * y * SQR((1 + z ^ 2))
END IF
z1 = -32.17405 / y ^ 2
v1 = z
w1 = 1 / y
END SUB
```

3

COMPUTATION OF THE FIRING RANGE

Introduction

The PC programs included in this chapter can be used to find the horizontal range of the projectile and other elements of a projectile trajectory when a projectile is launched with a given initial speed on condition that the departure angle and the ballistics coefficient are known.

The PC programs presented hereafter can be classified into three categories:

- The PC programs that use fixed BC that is constant for any trajectory of the given projectile.
- The PC programs that employ a BC that is a function of the projectile speed.
- The PC programs that use a BC that is a function of the departure angle.

The PC programs can be grouped in two other categories according to the exterior ballistics problems they solve:

- Finding the horizontal range;
- Finding the drop of a projectile (especially a bullet) at any range, or the elements of a projectile trajectory simultaneously in two points.

For all programs, we assume that the origin of the coordinates is at the sea level.

3.1 PC Program Rangec.Bas

The PC program Rangec.Bas can be used to determine the horizontal range of a projectile, or the trajectory elements at a given point when the projectile is fired with a given departure speed and departure angle.

The projectile is fired in the Standard Atmosphere. The firearm and the point of impact are at the same level over the sea.

The program estimates the coordinates of the trajectory vertex, as well as the time of flight, the terminal speed and the terminal angle.

The program finds the trajectory elements at a point on the trajectory for a given x-coordinate of that point. The value of x-coordinate of the point must be less than the range of the projectile.

The program uses a constant BC that corresponds to the Siacci function in use in the book and represents an appropriate ballistics coefficient (or a measured BC at a particular distance) that usually is given in the ballistics tables, or in the firearm handbooks.

Use of Rangec.Bas

Example 3.1

The bullet caliber 7.62mm is fired from a Russian AK47 with an initial speed 710ms. The BC is 5.918

   (a) Find the horizontal range if the departure angle is 0.6 degree. Assume that the flight of the bullet is in standard atmosphere, and the AK47 is at the sea level. Find as well the trajectory characteristics at the point with abscissa 300m.
   (b) Find the horizontal range if there is a range wind of 5m/s, as well as the cross deflection of the bullet due to a cross wind of 5m/s.
   (c) Find the horizontal range if the fire is at altitude 1000m over the sea level, if there is a range wind of 5m/s.

Solution

(a) **Input**: Y-coordinate of Firearm, 0; departure speed, 710; departure angle, 0.6; BC, 5.918; Integration step, 1; Abscissa x of a point on the trajectory, 300.

 Output: Range, 501m; Time, 1.19; Speed, 285m/s; Impact angle, -1.10178; Vertex, (290, 1.8).

 For the point with abscissa x = 300m we have: Corresponding ordinate y = 1.8m; time = 0.58s; speed = 391; angle = -0.0401.

(b) In the same way, we find (integration step 0.1); range, 504m; time, 1.19; Speed, 285m/s; Impact angle, -1.1004; Vertex, (292, 1.8), cross wind deflections, 2.43m.

(c) In the same way as in (a) we find that the horizontal range, at the altitude 1000m, is 523m.

Example 3.2

In Example 5, "Exterior Ballistics with Applications", (G. Klimi, p. 97-98, Xlibris 2008) for the bullet caliber 0.30 Ball M2 we found an average BC of 3.385.

Use BC = 3.385 to find the zero range of the bullet if the departure angle is 18 minutes (0.3 degree). Initial speed of the bullet is 853.44m/s. The atmosphere is standard and the shooting takes place at the sea level.

Note. Zero range is the point where the bullet intersects the line of sight. We do not consider the sight height of the gun.

Solution

Using RangeC.Bas

Input: Initial y-coordinate, 0; Departure angle, 0.30; Initial speed, 853.44; BC, 3.385; Integration step, 1.

Output: Zero range, 525m; Time, 0.82s; Impact speed, 483.17m/s; Impact angle, -0.43957 degree; Trajectory vertex location, (291m, 0.83m).

Exercise 3.3

Find the range of a projectile of mass 16.58kg fired with an initial speed of 580m/s at an angle 5 degree from a 107mm cannon. The form coefficient of the projectile is 0.594.

Find as well the coordinates of the trajectory vertex. The atmosphere is standard and the shooting takes place at the sea level

Answer. Horizontal range, 3516m; vertex, (1952m, 97m).

Exercise 3.4

The Siacci ballistics coefficient of 7.62mm M80 bullet (see section 1.1) is 3.182. The bullet is fired with a speed of 856.50m/s.

(a) Find the horizontal range of the bullet if the departure angle is 0.1398 degree. Find as well the y-coordinate of the bullet trajectory at the point with abscissa 200m. The atmosphere is standard.

(b) Find the range of the bullet fired at the same angle (0.1398) if it is a range wind of 6m/s directed against the flight. Find the cross deflection of the bullet due to a cross wind of 5m/s.

(c) Considering the data in question (a), find the range if the shooting takes place on a high mountain 1200 meters over the sea level. How high will the bullet pass over the center of the target located at the range 300meters?

(d) Find the horizontal range of the given bullet (at the sea level) for the same departure angle (0.1398 degree), if the bullet is fired with speed 800m/s.

(e) Use the PC program Anglec.Bas to find the departure angle needed to zero the gun at 300 meters.

Answer. (a) Horizontal range, 300m; the y-coordinate that corresponds to the horizontal distance 200 meters is 0.185m. (b) Horizontal range 299.57m, cross deflection, 0.26m. (c) Horizontal range, 306m; bullet passes 0.02m over the center of the target. (d) Horizontal range, 267m; the departure angle needed to zero the gun at 300 meters is 0.1613 degree.

QBasic PC Program PC Program RANGEC.BAS
Standard Atmosphere, Wind is Present

```
'FIND:    Range, Vertex of the Trajectory, etc.
          'Elements of Trajectory for a given abscissa x
'GIVEN:  Departure Angle, Ballistics Coefficient, Projectile Speed
'_____

'FIND: Range, Vertex of the Trajectory, etc.
'Elements of Trajectory for a given abscissa x
'GIVEN: Departure Angle, Ballistics Coefficient, Projectile Speed
'_____

'Control Data
'Input Data
'Input: Initial x-Coordinate = 0, Initial y-coordinate
'Input: Launching Angle: 6.20, Initial Speed = 885, Ballistics Coefficient
     =0.2389
'Input: Initial Time t0 = 0, Integration Step h0 =1
'Input: x-coordinate of a point on trajectory xc = 4530
'Results
'Range = 10000; Error in y-coordinate = -0.02972; Terminal Speed = 419.45
'Terminal Angle = -10.19; Coordinates of vertex (5634, 346.81)
'Abscissa of a Point on the trajectory = 4530,
'Corresponding values of x = 4350 are: y-coordinate =330.51; Speed = 639.07;
     time = 6.047; angle = 1.649859
'_____

'Functions, Subs

DECLARE SUB y1z1v1w1 (x, y, z, v, w, y1, z1, v1, w1, koef, ys, yy, wind)
DECLARE SUB InfHyres (x0, y0, z0, v0, w0, a, koef, xc1, yT, h0, wind, cw, vo)
DECLARE SUB NPxyzvw (nk, x, x0, y, y0, z, z0, v, v0, w, w0, h, h0, k, L, r, q)
DECLARE SUB NPkoef (k, L, r, q, h, y1, z1, v1, w1)
DECLARE SUB menu (cog, cof, xf, yf, xfu, yfu, t$)
DECLARE SUB c (koef)

'Variables
DIM m(4, 4), v(4)
rendi = 4
cog = 7: cof = 0
```

```
'Solution
CLS
fillimi:
menu cog, cof, 3, 10, 21, 70, "Initial Data"
InfHyres x0, y0, z0, v0, w0, a, koef, xc1, yT, h0, wind, cw, vo
c koef
f:
FOR nk = 1 TO rendi
NPxyzvw nk, x, x0, y, y0, z, z0, v, v0, w, w0, h, h0, k, L, r, q
y1z1v1w1 x, y, z, v, w, y1, z1, v1, w1, koef, ys, yy, wind
NPkoef k, L, r, q, h, y1, z1, v1, w1
m(nk, 1) = k: m(nk, 2) = L
m(nk, 3) = r: m(nk, 4) = q
NEXT nk

'Calculation
FOR i = 1 TO rendi
v(i) = 1 / 6 * (m(1, i) + 2 * m(2, i) + 2 * m(3, i) + m(4, i))
NEXT i

'New Data
x0 = x0 + h: y0 = y0 + v(1): z0 = z0 + v(2)
v0 = v0 + v(3): w0 = w0 + v(4)
IF y0 >= 256 THEN

IF ABS(z0) < .00001 OR ABS(z0) <= .0001 THEN
ymax = v0
xmax = x0 + wind * w0
END IF
END IF

IF y0 < 256 THEN
IF ABS(z0) < .0001 OR ABS(z0) < .001 THEN
ymax = v0
xmax = x0 + wind * w0
END IF
END IF

xab = x0 + wind * w0
```

```
IF (xab - xc1) <= .01 THEN
xc = xc1
yc = v0
tc = w0
ac = (180 / 3.141592654#) * ATN(z0)
vc = y0 / COS(ATN(z0))
END IF


IF v0 - yT <= .001 THEN
'Display Results
menu cog, cof, 2, 20, 22, 76, "RESULTS:"
LOCATE 3, 26: PRINT "Departure Angle       ="; a
LOCATE 4, 26: PRINT "Departure Speed       ="; vo
LOCATE 6, 26: PRINT "Horizontal Range  [m] = "; INT((x0 + wind * w0)
    * 100 + .5) / 100
LOCATE 7, 26: PRINT "Error in y-coord [m] ="; INT((v0 - yT) * 1000 + .5) / 1000
LOCATE 8, 26: PRINT "Time of Flight [s]   = "; INT((w0) * 100 + .5) / 100
LOCATE 9, 26: PRINT "Terminal Speed [m/s] = "; INT((y0 * (1 + z0 ^ 2) ^
    .5) * 100 + .5) / 100
LOCATE 10, 26: PRINT "Terminal Angle [Deg.] = "; ATN(z0) * 180 / 3.141593
LOCATE 11, 26: PRINT "Trajectory Vertex (xm, ym)   = "; "("; INT((xmax)
    * 100 + .5) / 100; ","; INT((ymax) * 100 + .5) / 100; ")"
LOCATE 12, 26: PRINT "Cross-Wind Deflection     = "; INT((cw * (w0 - x0
    / (vo * COS(a * 3.14159265# / 180)))) * 1000 + .5) / 1000
LOCATE 14, 23: PRINT "Abscissa of a point on trajectory:  X = "; INT((xc)
    * 100 + .5) / 100
LOCATE 15, 23: PRINT "Corresponding ordinate of x:      Y = "; INT((yc)
    * 1000 + .5) / 1000
LOCATE 16, 23: PRINT "Corresponding Time:             T = "; INT((tc) *
    100 + .5) / 100
LOCATE 17, 23: PRINT "Corresponding Speed:            V = "; INT((vc) *
    100 + .5) / 100
LOCATE 18, 23: PRINT "Corresponding Angle:            A = "; ac
LOCATE 19, 23: PRINT "Cross-Wind Deflection          D = "; INT((cw *
    (tc - xc1 / (vo * COS(a * 3.14159265# / 180)))) * 1000 + .5) / 1000
LOCATE 21, 24: PRINT "Ballistics Coefficient = "; koef
ELSE
GOTO f:
END IF
END
```

```
SUB c (koef)
koef = koef
END SUB

SUB InfHyres (x0, y0, z0, v0, w0, a, koef, xc1, yT, h0, wind, cw, vo)
LOCATE 5, 13: INPUT "y-coordinate of Firearm:          "; v0
LOCATE 6, 13: INPUT "Departure SPEED [m/s]:           "; y0
LOCATE 7, 13: INPUT "Departure ANGLE [Degree]:        "; z0
LOCATE 8, 13: INPUT "Range Wind:                   "; wind
LOCATE 9, 13: INPUT "Cross Wind:                   "; cw
LOCATE 10, 13: INPUT "Ballistics Coefficient:        "; koef
LOCATE 11, 13: INPUT "X coordinate of a point on Trajectory [m]  = "; xc1
LOCATE 12, 13: INPUT "Integration Step: 10; 1; 0.5; 0.1  "; h0
yT = v0
vo = y0
a = z0
y0 = SQR(vo ^ 2 + wind ^ 2 - 2 * vo * wind * COS(a * 3.141592654# / 180))
y0 = y0 * COS(a * 3.141592654# / 180)
z0 = TAN(a * 3.141592654# / 180)
z0 = z0 / (1 - wind / (vo * COS(a * 3.141592654# / 180)))
CLS
END SUB

SUB menu (cog, cof, xf, yf, xfu, yfu, t$)
COLOR cog, cof
LOCATE xf - 1, yf: PRINT t$
LOCATE xf, yf: PRINT "É" + STRING$(yfu - yf, 205) + "»";
FOR i = xf + 1 TO xfu
LOCATE i, yf: PRINT "°" + SPACE$(yfu - yf) + "°";
NEXT
LOCATE xfu + 1, yf: PRINT "È" + STRING$(yfu - yf, 205) + "¼";
END SUB

SUB NPkoef (k, L, r, q, h, y1, z1, v1, w1)
k = h * y1: L = h * z1
```

```
r = h * v1: q = h * w1
END SUB


SUB NPxyzvw (nk, x, x0, y, y0, z, z0, v, v0, w, w0, h, h0, k, L, r, q)
IF nk = 1 THEN
x = x0: y = y0: z = z0
v = v0: w = w0: h = h0
GOTO fund:
END IF


IF nk = 2 OR nk = 3 THEN
x = x0 + (.5 * h): y = y0 + (.5 * k)
z = z0 + (.5 * L): v = v0 + (.5 * r)
w = w0 + (.5 * q)
GOTO fund:
END IF


IF nk = 4 THEN
x = x0 + h: y = y0 + k: z = z0 + L
v = v0 + r: w = w0 + q
END IF
fund:
END SUB


SUB y1z1v1w1 (x, y, z, v, w, y1, z1, v1, w1, koef, ys, yy, wind)
yy = y * SQR(1 + z ^ 2)
IF yy > 256! THEN
y1 = -1 * koef * ((289.08 - .006328 * v) / 289.08) ^ 4.4 * (yy - 240) / (3 * yy)
ELSE
y1 = -1 * koef * ((289.08 - .006328 * v) / 289.08) ^ 4.4 * .0001212 * yy ^ 2 / yy
END IF
z1 = -9.80665 / y ^ 2
v1 = z
w1 = 1 / y
END SUB
```

3.2 PC Program RCPoint.Bas, Bullet Drop, Bullet Path

The PC program RCPoint.Bas can be used to estimate the elements of a projectile trajectory simultaneously at any two points. The departure speed and the departure angle are given. It is known as well the ballistics coefficient of the projectile.

The RCPoint.Bas can be used to find the drop of a bullet fired horizontally (departure angle equal to zero degree) at any two points on the trajectory. The firearm can be at the sea level or at any altitude over the sea. The atmosphere is standard. The wind can be present or not.

Use of RCPoint.Bas

Example 3.5 Bullet Drop, Angle of Sight

The Siacci ballistics coefficient of 7.62mm M80 bullet (see section 1.1) is 3.182. The bullet is fired with a speed of 856.50m/s. Use the PC program RCpoint.Bas. The atmosphere is standard.

(a) Find the drop of the bullet at the points respectively with abscissa 300m and 500m if the bullet is fired horizontally from the gun located at the sea level.

    Find as well the departure angle and the angle of sight to zero the gun at 300 meters, and 500meters if the sight height is 1.5 inches (0.0381m).

(b) Find the drop of the bullet at the points respectively with abscissa 300 and 500 if the bullet is fired horizontally from the gun located at a mountain 1200 meters over the sea level.

Solution

(a) **Input**: Altitude of Firearm, 0; departure speed, 856.50; departure angle, 0; x-coordinate of a point, 300; X-coordinate of another point, 500, BC =3.182.

    Output: At the point with abscissa 300, the drop is -0.731m; at the point with abscissa 500, the drop is -2.361.

Departure Angle

Using (1.9.22), we find the departure angle

$$\alpha_0 = \frac{\overline{y}}{x} = \frac{0.731}{300} = 0.002437 = 0.1396° = 8.38 MOA$$

Using (1.7.7), we find that the angle of sight is

$$\alpha_{S0} = 60 \cdot \alpha_0 + \frac{10800}{\pi} \frac{h_S}{D_0} = 60 \cdot (0.002437) + \frac{10800}{\pi} \cdot \frac{0.0381}{300} = 8.81 MOA.$$

In the same way, we find that to zero the rifle at 500 meters the departure angle is

$$\alpha_0 = 0.27055° = 16.23 MOA.$$

The corresponding angle of sight is

$$\alpha_{S0} = 16.49 MOA$$

(b) In the same way as in (a), but inputting 1200 for the altitude of fire, we find that the drops at 300 and 500 meters are respectively -0.714m and -2.26m.

The departure angle needed to zero the gun at the horizontal range 300 meters and the sight angle are respectively 8.18 MOA and 8.62MOA.

Exercise 3.6

Employ RCPoint.Bas and use the departure angle (8.38MOA) found in example 3.6 to find the height of the trajectory respectively at the points with abscissa 300 and 500 meters, if the 7.62mm M80 bullet is fired from the muzzle of the rifle located 1.6 meters over the sea level.
Answer: 0m and -1.142m.

Example 3.7 Bullet Drop, Bullet Path

A bullet caliber 7.62mm M852 HPBT is fired from a M14 rifle with an initial speed 807.70m/s. The ballistics coefficient of the bullet is 2.945. Use the PC Program RCPoint.Bas.

(a) Find the drop of the bullet at the trajectory point that corresponds to the horizontal range 300.
(b) Use the result in (a) to find the departure angle (aiming angle) that zeros the rifle at the horizontal range 300m, as well as the y-coordinate of the trajectory that corresponds to the horizontal range 200m.
(c) Find the bullet path respectively at the horizontal ranges 100, 200, and 300 meters.
(d) Find the angle of sight (The angle that the departure line forms with the LOS).

Note: The bullet path is called the distance of a point on the trajectory from the line of sight (LOS).

Solution

(a) **Input**: Altitude of firearm, 0; departure speed, 807.70; departure angle, 0; abscissa x of a trajectory point, 200; abscissa x of another trajectory point, 300; integration step, 0.5; BC, 2.945.
    Output: projectile drop at 300m is -0.814.
(b) The departure angle is

$$\alpha_0 = \frac{Drop}{x} = \frac{0.814}{300}\frac{180}{\pi} = 0.155463° = 9.328 MOA .$$

Using RCPoint.Bas and the above departure angle, we find that the y-coordinate of the point on the trajectory that corresponds to abscissa 200 is 0.204m.

(c) At 200 meters the LOS is

$$y_L = 0.0381 \frac{200}{300} = 0.0254m$$

above the horizontal line. The path of the bullet is

$$Path = 0.204 - 0.0254 = 0.179m.$$

(d) Using the formula (1.7.7) we find that the angle of sight is

$$\alpha_{S0} = \alpha_0 + \frac{h_S}{D_0} \frac{10800}{\pi} = 9.328 + \frac{0.0381}{300} \cdot \frac{10800}{\pi} = 9.764 MOA.$$

PC Program RCPoint.Bas
Standard Atmosphere, Wind Present

'FIND: Elements of the Projectile Trajectory simultaneously at two points
'GIVEN: Departure speed, Departure Angle, Ballistics Coefficient
'_____

'Control Data
'Input Data
'Projectile Drop
'Input: Altitude of Firearm = 0;
'Input: Departure Speed [m/s]   = 856.5,
'Input: Departure Angle [Degree] = 0
'Input: Abscissa x of a point [m] = 300
'Input: Abscissa X of another point= 500
'Input: Integration Step   = 0.5
'Input: Cross wind speed [m/s]   = 5
'Ballistics Coefficient   3.182

'OUTPUT
'Altitude of Firearm
'Abscissa [m]:   x = 300
'Corresponding Ordinate: y = -0.731 (Drop)
'Corresponding Speed: v = 641
'Time of Flight [s]:   t = 0.405
'Corresponding Angle   a = -.3089 degree
'Cross Deviation   z = 0.27
'Abscissa   X = 500
'Corresponding Ordinate Y = -2.36 (Drop)
'Corresponding Speed   V = 518
'Time   T = 0.75
'Corresponding Angle   A = -0.6493
'Cross Deviation   Z = 0.84
'_____

'Functions, Subs
DECLARE SUB y1z1v1w1 (x, y, z, v, w, y1, z1, v1, w1, koef, ys, yy, wind)
DECLARE SUB InfHyres (x0, y0, z0, v0, w0, a, koef, xc1, h0, vo, xx1, alt,
    xax, aa, vv, cw, wind)
DECLARE SUB NPxyzvw (nk, x, x0, y, y0, z, z0, v, v0, w, w0, h, h0, k, L, r, q)
DECLARE SUB NPkoef (k, L, r, q, h, y1, z1, v1, w1)

```
DECLARE SUB menu (cog, cof, xf, yf, xfu, yfu, t$)
DECLARE SUB c (koef, y0, a)

'Variables
DIM m(4, 4), v(4)
rendi = 4
cog = 7: cof = 0

'Solution
CLS

fillimi:
menu cog, cof, 3, 10, 21, 70, "Initial Data"
InfHyres x0, y0, z0, v0, w0, a, koef, xc1, h0, vo, xx1, alt, xax, aa, vv, cw, wind
c koef, y0, a

f:
FOR nk = 1 TO rendi
NPxyzvw nk, x, x0, y, y0, z, z0, v, v0, w, w0, h, h0, k, L, r, q
y1z1v1w1 x, y, z, v, w, y1, z1, v1, w1, koef, ys, yy, wind
NPkoef k, L, r, q, h, y1, z1, v1, w1
m(nk, 1) = k: m(nk, 2) = L
m(nk, 3) = r: m(nk, 4) = q
NEXT nk

'Calculation
FOR i = 1 TO rendi
v(i) = 1 / 6 * (m(1, i) + 2 * m(2, i) + 2 * m(3, i) + m(4, i))
NEXT i

'New Data
x0 = x0 + h: y0 = y0 + v(1): z0 = z0 + v(2)
v0 = v0 + v(3): w0 = w0 + v(4)
xxx = x0 + wind * w0
IF (xxx - xx1) <= .001 THEN
xc = xxx
yc = v0
tc = w0
ac = 180 * ATN(z0) / 3.141592654#
```

```
vc = y0 * (1 + z0 ^ 2) ^ .5
END IF


xxT = x0 + wind * w0
IF (xxT - xax) <= .001 THEN
tt = w0
xt = xxT
yt = v0
at = 180 * ATN(z0) / 3.141592654#
vt = y0 * (1 + z0 ^ 2) ^ .5
END IF


ytt = yt
IF ABS(v0 - ytt) >= .1 THEN
'Display Results
menu cog, cof, 2, 20, 22, 76, "RESULTS:"
LOCATE 5, 23: PRINT "Altitude of Firearm        yo = "; vo
LOCATE 7, 23: PRINT "Point with abscissa:       x = "; INT((xc) + .5)
LOCATE 8, 23: PRINT "Corresponding Ordinate:  y="; INT((yc) * 1000 + .5) / 1000
LOCATE 9, 23: PRINT "Corresponding Speed:       v = "; INT(vc + .5)
LOCATE 10, 23: PRINT "Corresponding Time:      t = "; INT((tc) * 100 + .5) / 100
LOCATE 11, 23: PRINT "Corresponding Angle:       a = "; ac
LOCATE 12, 23: PRINT "Cross Deflection[m]:       z = "; INT((cw * (tc - xc
    / (vv * COS(a * 3.14159265# / 180)))) * 100 + .5) / 100
LOCATE 14, 23: PRINT "Point with abscissa   X = "; INT((xt) + .5)
LOCATE 15, 23: PRINT "Corresponding Ordinate Y ="; INT((yt) * 1000 + .5) / 1000
LOCATE 16, 23: PRINT "Corresponding Speed:  V = "; INT((vt) + .5)
LOCATE 17, 23: PRINT "Time of Flight:       T = "; INT((tt) * 100 + .5) / 100
LOCATE 18, 23: PRINT "Corresponding Angle:  A = "; at
LOCATE 19, 23: PRINT "Cross Deflection[m]:  Z = "; INT((cw * (tt - xt / (vv
    * COS(a * 3.14159265# / 180)))) * 100 + .5) / 100
LOCATE 21, 23: PRINT "Ballistics Coefficient:  = "; koef
ELSE
GOTO f:
END IF
END


SUB c (koef, y0, a)
```

```
koef = koef
END SUB


SUB InfHyres (x0, y0, z0, v0, w0, a, koef, xc1, h0, vo, xx1, alt, xax, aa, vv,
    cw, wind)
LOCATE 5, 13: INPUT "Altitude of the FIREARM [m]         :"; v0
LOCATE 6, 13: INPUT "Deparure SPEED [m/s]              :"; y0
LOCATE 7, 13: INPUT "Departura ANGLE [Degree]           :"; z0
LOCATE 8, 13: INPUT "Abscissa x of a point on trajectory   :"; xx1
LOCATE 9, 13: INPUT "Abscissa X of another trajectory point :"; xax
LOCATE 10, 13: INPUT "Range wind                  :"; wind
LOCATE 11, 13: INPUT "Cross-wind                 :"; cw
LOCATE 12, 13: INPUT "Integration Step (10, 1, 0.5)     :"; h0
LOCATE 13, 13: INPUT "Ballistics Coefficient         :"; koef
vo = v0
alt = v0
a = z0
aa = z0
vv = y0
xc1 = 0
y0 = SQR(vv ^ 2 + wind ^ 2 - 2 * vv * wind * COS(a * 3.141592654# / 180))
y0 = y0 * COS(a * 3.141592654# / 180)
z0 = TAN(a * 3.141592654# / 180)
z0 = z0 / (1 - wind / (vv * COS(a * 3.141592654# / 180)))
CLS
END SUB


SUB menu (cog, cof, xf, yf, xfu, yfu, t$)
COLOR cog, cof
LOCATE xf - 1, yf: PRINT t$
LOCATE xf, yf: PRINT "É" + STRING$(yfu - yf, 205) + "»";
FOR i = xf + 1 TO xfu
LOCATE i, yf: PRINT "º" + SPACE$(yfu - yf) + "º";
NEXT
LOCATE xfu + 1, yf: PRINT "È" + STRING$(yfu - yf, 205) + "¼";
END SUB


SUB NPkoef (k, L, r, q, h, y1, z1, v1, w1)
```

```
k = h * y1: L = h * z1
r = h * v1: q = h * w1
END SUB


SUB NPxyzvw (nk, x, x0, y, y0, z, z0, v, v0, w, w0, h, h0, k, L, r, q)
IF nk = 1 THEN
x = x0: y = y0: z = z0
v = v0: w = w0: h = h0
GOTO fund:
END IF


IF nk = 2 OR nk = 3 THEN
x = x0 + (.5 * h): y = y0 + (.5 * k)
z = z0 + (.5 * L): v = v0 + (.5 * r)
w = w0 + (.5 * q)
GOTO fund:
END IF


IF nk = 4 THEN
x = x0 + h: y = y0 + k: z = z0 + L
v = v0 + r: w = w0 + q
END IF
fund:
END SUB


SUB y1z1v1w1 (x, y, z, v, w, y1, z1, v1, w1, koef, ys, yy, wind)
yy = y * SQR(1 + z ^ 2)


IF yy > 256! THEN
y1 = -1 * koef * ((289.08 - .006328 * v) / 289.08) ^ 4.4 * (yy - 240) / (3 * yy)
ELSE
y1 = -1 * koef * ((289.08 - .006328 * v) / 289.08) ^ 4.4 * .0001212 * yy ^ 2 / yy
END IF
z1 = -9.80665 / y ^ 2
v1 = z
w1 = 1 / y
END SUB
```

3.3 PC Program RNATO762.Bas, 7.62mm NATO Bullet

The PC program RNATO762.Bas can be used to estimate the horizontal range of a 7.62mm M852 HPBT bullet as well as the elements of the trajectory at a point, whose abscissa is known.

Instead of RNATO762.Bas we can use the program Rangec.Bas to estimate the horizontal range of a 7.62mm M852 HPBT bullet, considering a fixed ballistics coefficient equal to 2.945 (ref. section 1.1).

Example 3.8 Bullet Drop, Angle of sight

A bullet caliber 7.62mm M852 HPBT is fired from a M14 rifle at an angle 0.275 degree with an initial speed 807.72m/s. The atmosphere is standard, but there is a range wind of 10ms in opposite direction of flight, as well as a cross wind of 5m/s.

Use the PC Program RNATO762.Bas to find the horizontal range and the elements of the trajectory at the point with abscissa 300 meters.

Solution

Input: y-coordinate of gun, 0; departure speed, 807.72m/s; departure angle, 0.275; range wind, -10; cross wind, 5m/s; x-coordinate of a point on trajectory, 300; integration step, 0.5.

Output: Horizontal range 468m; time of flight, 0.73s; terminal speed, 521.67; terminal angle, -0.3336, vertex (256, m0.65m); cross deflection, 0.7m.

At the point with abscissa 300 meters, we have the following values:

Y-coordinate of the point, 0.624m; time, 0; speed, 619m/s; angle, 0.06805; cross deflection, 0.28m.

PC Program RNATO762.Bas
7.62mm M852 HPBT Bullet

```
'FIND:  Range, Vertex of the Trajectory, etc.
          'Elements of Trajectory for a given abscissa x
'GIVEN: Launching Angle, Ballistics Coefficient, Projectile Speed
'_____

'Control Data
'Input Data
'Input: y-coordinate of gun, 0
'Input: Departure Angle: 0.298695, Initial Speed = 807.72
'Range wind, 0; cross wind, 0;
'Integration Step h0 = 0.5
'Input: x-coordinate of a point on trajectory xc = 500
'Output: Range = 500; Error in y-coordinate = 0; Terminal Speed = 502
'Terminal Angle = -0.422145; Coordinates of vertex (273.5, 0.76)
'_____

'Functions, Subs
DECLARE SUB y1z1v1w1 (x, y, z, v, w, y1, z1, v1, w1, koef, ys, yy, wind)
DECLARE SUB InfHyres (x0, y0, z0, v0, w0, a, koef, xc1, yT, h0, wind, cw, vo)
DECLARE SUB NPxyzvw (nk, x, x0, y, y0, z, z0, v, v0, w, w0, h, h0, k, L, r, q)
DECLARE SUB NPkoef (k, L, r, q, h, y1, z1, v1, w1)
DECLARE SUB menu (cog, cof, xf, yf, xfu, yfu, t$)
DECLARE SUB c (koef, y0)

'Variables
DIM m(4, 4), v(4)
rendi = 4
cog = 7: cof = 0

'Solution
CLS
fillimi:
menu cog, cof, 3, 10, 21, 70, "Initial Data"
InfHyres x0, y0, z0, v0, w0, a, koef, xc1, yT, h0, wind, cw, vo
c koef, y0
```

```
f:
FOR nk = 1 TO rendi
NPxyzvw nk, x, x0, y, y0, z, z0, v, v0, w, w0, h, h0, k, L, r, q
y1z1v1w1 x, y, z, v, w, y1, z1, v1, w1, koef, ys, yy, wind
NPkoef k, L, r, q, h, y1, z1, v1, w1
m(nk, 1) = k: m(nk, 2) = L
m(nk, 3) = r: m(nk, 4) = q
NEXT nk

'Calculation
FOR i = 1 TO rendi
v(i) = 1 / 6 * (m(1, i) + 2 * m(2, i) + 2 * m(3, i) + m(4, i))
NEXT i

'New Data
x0 = x0 + h: y0 = y0 + v(1): z0 = z0 + v(2)
v0 = v0 + v(3): w0 = w0 + v(4)
IF y0 >= 256 THEN
IF ABS(z0) < .00001 OR ABS(z0) <= .0001 THEN
ymax = v0
xmax = x0 + wind * w0
END IF
END IF
IF y0 < 256 THEN
IF ABS(z0) < .0001 OR ABS(z0) < .001 THEN
ymax = v0
xmax = x0 + wind * w0
END IF
END IF

xab = x0 + wind * w0
IF (xab - xc1) <= .001 THEN
xc = xab
yc = v0
tc = w0
```

```
ac = (180 / 3.141592654#) * ATN(z0)
vc = y0 / COS(ATN(z0))
END IF


IF v0 - yT <= .001 THEN


'Display Results
menu cog, cof, 2, 20, 22, 76, "RESULTS:"
LOCATE 3, 26: PRINT "Departure Angle       ="; a
LOCATE 4, 26: PRINT "Departure Speed       ="; vo
LOCATE 6, 26: PRINT "Horizontal Range  [m] = "; INT((x0 + wind * w0) + .5)
LOCATE 7, 26: PRINT "Error in y-coord [m]  ="; INT((v0 - yT) * 1000 + .5) / 1000
LOCATE 8, 26: PRINT "Time of Flight [s]    = "; INT((w0) * 100 + .5) / 100
LOCATE 9, 26: PRINT "Terminal Speed    = "; INT((y0 * (1 + z0 ^ 2) ^ .5)
    * 100 + .5) / 100
LOCATE 10, 26: PRINT "Terminal Angle [Deg.] = "; ATN(z0) * 180 / 3.141593
LOCATE 11, 26: PRINT "Trajectory Vertex (xm, ym)  = "; "("; INT((xmax)
    * 100 + .5) / 100; ","; INT((ymax) * 100 + .5) / 100; ")"
LOCATE 12, 26: PRINT "Cross-Wind Deflection       = "; INT((cw * (w0 - x0
    / (vo * COS(a * 3.14159265# / 180)))) * 100 + .5) / 100
LOCATE 14, 23: PRINT "Abscissa of a point on trajectory:  X = "; INT((xc) + .5)
LOCATE 15, 23: PRINT "Corresponding ordinate of x:   Y = "; INT((yc) *
    1000 + .5) / 1000
LOCATE 16, 23: PRINT "Corresponding Time:      T ="; INT((tc) * 100 + .5) / 100
LOCATE 17, 23: PRINT "Corresponding Speed:      V ="; INT((vc) * 100 + .5) / 100
LOCATE 18, 23: PRINT "Corresponding Angle:             A = "; ac
LOCATE 19, 23: PRINT "Cross-Wind Deflection          D = "; INT((cw *
    (tc - xc1 / (vo * COS(a * 3.14159265# / 180)))) * 100 + .5) / 100
LOCATE 21, 24: PRINT "Ballistics Coefficient = "; koef
ELSE
GOTO f:
END IF
END


SUB c (koef, y0)
koef = 1.0563 * (1.2112105619# + .01289224624# * y0 - .0000261226# * y0
    ^ 2 + .000000015573# * y0 ^ 3)
END SUB
```

```
SUB InfHyres (x0, y0, z0, v0, w0, a, koef, xc1, yT, h0, wind, cw, vo)
LOCATE 5, 13: INPUT "y-coordinate of Firearm:        "; v0
LOCATE 6, 13: INPUT "Departure SPEED [m/s]:          "; y0
LOCATE 7, 13: INPUT "Departure ANGLE [Degree]:         "; z0
LOCATE 8, 13: INPUT "Range Wind:                  "; wind
LOCATE 9, 13: INPUT "Cross Wind:                 "; cw
LOCATE 10, 13: INPUT "X coordinate of a point on Trajectory [m]  = "; xc1
LOCATE 11, 13: INPUT "Integration Step: 10; 1; 0.5; 0.1  "; h0
yT = v0
vo = y0
a = z0
y0 = SQR(vo ^ 2 + wind ^ 2 - 2 * vo * wind * COS(a * 3.141592654# / 180))
y0 = y0 * COS(a * 3.141592654# / 180)
z0 = TAN(a * 3.141592654# / 180)
z0 = z0 / (1 - wind / (vo * COS(a * 3.141592654# / 180)))
CLS
END SUB


SUB menu (cog, cof, xf, yf, xfu, yfu, t$)
COLOR cog, cof
LOCATE xf - 1, yf: PRINT t$
LOCATE xf, yf: PRINT "É" + STRING$(yfu - yf, 205) + "»";
FOR i = xf + 1 TO xfu
LOCATE i, yf: PRINT """ + SPACE$(yfu - yf) + """;
NEXT
LOCATE xfu + 1, yf: PRINT "È" + STRING$(yfu - yf, 205) + "¼";
END SUB


SUB NPkoef (k, L, r, q, h, y1, z1, v1, w1)
k = h * y1: L = h * z1
r = h * v1: q = h * w1
END SUB


SUB NPxyzvw (nk, x, x0, y, y0, z, z0, v, v0, w, w0, h, h0, k, L, r, q)
IF nk = 1 THEN
x = x0: y = y0: z = z0
v = v0: w = w0: h = h0
```

```
GOTO fund:
END IF


IF nk = 2 OR nk = 3 THEN
x = x0 + (.5 * h): y = y0 + (.5 * k)
z = z0 + (.5 * L): v = v0 + (.5 * r)
w = w0 + (.5 * q)
GOTO fund:
END IF


IF nk = 4 THEN
x = x0 + h: y = y0 + k: z = z0 + L
v = v0 + r: w = w0 + q
END IF
fund:
END SUB


SUB y1z1v1w1 (x, y, z, v, w, y1, z1, v1, w1, koef, ys, yy, wind)
yy = y * SQR(1 + z ^ 2)
IF yy > 256! THEN
y1 = -1 * koef * ((289.08 - .006328 * v) / 289.08) ^ 4.4 * (yy - 240) / (3 * yy)
ELSE
y1 = -1 * koef * ((289.08 - .006328 * v) / 289.08) ^ 4.4 * .0001212 * yy ^
    2 / yy
END IF
z1 = -9.80665 / y ^ 2
v1 = z
w1 = 1 / y
END SUB
```

3.4 PC Program RCNATO762.Bas, 7.62mm NATO Bullet

The PC program RCNATO762.Bas is the PC program RCPoint.Bas, adapted to estimate only the elements of the trajectory of the NATO 7.62mm M855 HPBT bullet, as well as the bullet drop.

Instead of RCNATO762 we can use the program RCPoint.Bas to estimate the drop and the other the elements of the NATO7.62mm bullet, considering a fixed ballistics coefficient equal to 2.945 (ref. section 1.1).

The PC program RCNATO762.Bas is the PC program RCPoint.Bas, adapted to estimate only the elements of the trajectory of the NATO 7.62mm M855 HPBT bullet, as well as the bullet drop.

Instead of RCNATO762 we can use the program RCPoint.Bas to estimate the drop and the other the elements of the NATO7.62mm bullet, considering a fixed ballistics coefficient equal to 2.945 (ref. section 1.1).

Use of RCNATO762.Bas

Example 3.9 Bullet Drop, Angle of sight

A bullet caliber 7.62mm M852 HPBT is fired from a M14 rifle with an initial speed 807.70m/s. Use the PC Program RNATO762.Bas.

(a) Find the drop of the bullet at the trajectory point that corresponds to the horizontal range 500m. The rifle is at the sea level. The atmospheric characteristics at the sea level are standard. Find as well the y-coordinate of the point of trajectory that corresponds to the horizontal range 400m.

(b) Shooting takes place on a mountain 1200m over the sea level, horizontal range 500 meters. The characteristics of atmosphere at the sea level are standard.

Find the drop of the bullet at the trajectory point that corresponds to the horizontal range 500m. Find as well the departure angle and the angle of sight needed to zero the rifle at 500 meters.

(c) Find the projectile drop at 500 meters if there is a range wind of 10m/s. The shooting takes place at the sea level.

(d) Find the projectile drop at 500 meters if there is a range wind of 10m/s. The shooting takes place on a mountain at 1200 meters over the sea level.

(e) Use the departure angle found in (b) and the firing data given in (b) to verify that indeed the departure angle zeros the rifle at 500 meters.

(f) Use the RCPoint.Bas to give answer to question presented in (b). Use BC= 2.945.

Solution

(a) **Input**: Departure speed, 807.72; departure angle, 0; Abscissa x of a point on trajectory, 400; Abscissa x of another point on trajectory, 500; integration step, 0.5.

   Results. Abscissa 400m, drop is -1.55m. Abscissa, 500m, drop is -2.606.

(b) Inputting the altitude of the firearm, 1200m, the x-coordinate 500m, we find that the drop is -2.502.

The departure angle to zero the firearm at 500m is

$$\alpha_0 = \frac{Drop}{x} = \frac{2.502}{500}\frac{180}{\pi} = 0.28671° = 17.20 MOA \cdot$$

The angle of sight is

$$\alpha_{s0} = \alpha_0 + \frac{h_s}{D_0}\frac{10800}{\pi} = 17.20 + \frac{0.0381}{500} \cdot \frac{10800}{\pi} = 17.46 MOA \cdot$$

Note. The angle of departure that zeros the rifle at 500m at the sea level is

$$\alpha_0 = \frac{Drop}{x} = \frac{2.606}{500}\frac{180}{\pi} = 0.29863° = 17.92 MOA .$$

(c) The drop is -2.574m. (d) The drop is -2.473m. (e) Input altitude of firearm, 1200m, departure angle, 0.28671, etc. We find that the y-coordinate at 500 meters is zero, i.e. the departure angle we find in (b) indeed zeros the rifle at horizontal range 500m. (f) The drop is -2.502.

Comparing the answers for (b) and (f) we can notice that practically there is no difference in the drop of projectile estimated using any of the two PC programs.

PC Program RCNATO762.Bas
7.62mm M852, HPBT NATO Bullet, 168 grain
Standard Atmosphere, Wind is Present

'FIND: Elements of the Projectile Trajectory simultaneously at two points
'GIVEN: Departure Angle, initial speed, Ballistics Coefficient
           'Standard Atmosphere, Wind present

'_____

'Control Data
'Input Data
'Estimate Projectile Drop at 300m
'Input: Altitude of Firearm = 0;
'Input: Departure Speed [m/s] = 807.72,
'Input: Departure Angle [Degree] = 0
'Input: Abscissa x of a point [m] = 300
'Input: Abscissa X of another point= 300
'Input: Integration Step = 0.5
'Input: Cross wind speed [m/s] = 5

'OUTPUT
'Altitude of Firearm
'Abscissa [m]:   x = 300
'Corresponding Ordinate: y = -0.8143 (Drop)
'Corresponding Speed: v = 614
'Time of Flight [s]:   t = 0.426
'Corresponding Angle   a = -.34234 degree
'Cross Deviation   z = 0.274
'Ballistics Coefficient 2.945
'_____

'Functions, Subs
DECLARE SUB y1z1v1w1 (x, y, z, v, w, y1, z1, v1, w1, koef, ys, yy, wind)
DECLARE SUB InfHyres (x0, y0, z0, v0, w0, a, koef, xc1, h0, vo, xx1, alt,
      xax, aa, vv, cw, wind)
DECLARE SUB NPxyzvw (nk, x, x0, y, y0, z, z0, v, v0, w, w0, h, h0, k, L, r, q)
DECLARE SUB NPkoef (k, L, r, q, h, y1, z1, v1, w1)
DECLARE SUB menu (cog, cof, xf, yf, xfu, yfu, t$)
DECLARE SUB c (koef, y0, a)

```
'Variables
DIM m(4, 4), v(4)
rendi = 4
cog = 7: cof = 0

'Solution
CLS
fillimi:
menu cog, cof, 3, 10, 21, 70, "Initial Data"
InfHyres x0, y0, z0, v0, w0, a, koef, xc1, h0, vo, xx1, alt, xax, aa, vv, cw, wind
c koef, y0, a

f:
FOR nk = 1 TO rendi
NPxyzvw nk, x, x0, y, y0, z, z0, v, v0, w, w0, h, h0, k, L, r, q
y1z1v1w1 x, y, z, v, w, y1, z1, v1, w1, koef, ys, yy, wind
NPkoef k, L, r, q, h, y1, z1, v1, w1
m(nk, 1) = k: m(nk, 2) = L
m(nk, 3) = r: m(nk, 4) = q
NEXT nk

'Calculation
FOR i = 1 TO rendi
v(i) = 1 / 6 * (m(1, i) + 2 * m(2, i) + 2 * m(3, i) + m(4, i))
NEXT i

'New Data
x0 = x0 + h: y0 = y0 + v(1): z0 = z0 + v(2)
v0 = v0 + v(3): w0 = w0 + v(4)
xxx = x0 + wind * w0
IF (xxx - xx1) <= .001 THEN
xc = xxx
yc = v0
tc = w0
ac = 180 * ATN(z0) / 3.141592654#
vc = y0 * (1 + z0 ^ 2) ^ .5
zc = cw * (tc - xc / (vv * COS(aa)))
END IF
```

```
xxT = x0 + wind * w0
IF (xxT - xax) <= .001 THEN
tt = w0
xt = xxT
yt = v0
at = 180 * ATN(z0) / 3.141592654#
vt = y0 * (1 + z0 ^ 2) ^ .5
zt = cw * (w0 - x0 / (vv * COS(aa)))
END IF

ytt = yt
IF ABS(v0 - ytt) >= .1 THEN
'Display Results
menu cog, cof, 2, 20, 22, 76, "RESULTS:"
LOCATE 5, 23: PRINT "Altitude of Firearm          yo = "; vo
LOCATE 7, 23: PRINT "Point with abscissa:          x = "; INT((xc) + .5)
LOCATE 8, 23: PRINT "Corresponding Ordinate: y="; INT((yc) * 1000 + .5) / 1000
LOCATE 9, 23: PRINT "Corresponding Speed:      v = "; INT(vc + .5)
LOCATE 10, 23: PRINT "Corresponding Time:      t ="; INT((tc) * 100 + .5) / 100
LOCATE 11, 23: PRINT "Corresponding Angle:      a = "; ac
LOCATE 12, 23: PRINT "Cross Deflection[m]:      z = "; INT((zc) * 100 + .5) / 100
LOCATE 14, 23: PRINT "Point with abscissa          X = "; INT((xt) + .5)
LOCATE 15, 23: PRINT "Corresponding Ordinate Y="; INT((yt) * 1000 + .5) / 1000
LOCATE 16, 23: PRINT "Corresponding Speed:      V = "; INT((vt) + .5)
LOCATE 17, 23: PRINT "Time of Flight:          T ="; INT((tt) * 100 + .5) / 100
LOCATE 18, 23: PRINT "Corresponding Angle:      A = "; at
LOCATE 19, 23: PRINT "Cross Deflection[m]:      Z="; INT((zt) * 100 + .5) / 100
LOCATE 21, 23: PRINT "Ballistics Coefficient:      BC = "; koef
ELSE
GOTO f:
END IF
END

SUB c (koef, y0, a)
koef = 1.0563 * (1.2112105619# + .01289224624# * y0 - .0000261226# * y0
    ^ 2 + .000000015573# * y0 ^ 3)
END SUB
```

```
SUB InfHyres (x0, y0, z0, v0, w0, a, koef, xc1, h0, vo, xx1, alt, xax, aa, vv, cw, wind)
LOCATE 5, 13: INPUT "Altitude of the FIREARM [m]          :"; v0
LOCATE 6, 13: INPUT "Departure SPEED [m/s]               :"; y0
LOCATE 7, 13: INPUT "Departure ANGLE [Degree]            :"; z0
LOCATE 8, 13: INPUT "Abscissa x of a point on trajectory  :"; xx1
LOCATE 9, 13: INPUT "Abscissa X of another trajectory point :"; xax
LOCATE 10, 13: INPUT "Range wind                         :"; wind
LOCATE 11, 13: INPUT "Cross-wind                         :"; cw
LOCATE 12, 13: INPUT "Integration Step (10, 1, 0.5)       :"; h0
vo = v0
alt = v0
a = z0
aa = z0
vv = y0
xc1 = 0
y0 = SQR(vv ^ 2 + wind ^ 2 - 2 * vv * wind * COS(a * 3.141592654# / 180))
y0 = y0 * COS(a * 3.141592654# / 180)
z0 = TAN(a * 3.141592654# / 180)
z0 = z0 / (1 - wind / (vv * COS(a * 3.141592654# / 180)))
CLS
END SUB


SUB menu (cog, cof, xf, yf, xfu, yfu, t$)
COLOR cog, cof
LOCATE xf - 1, yf: PRINT t$

LOCATE xf, yf: PRINT "É" + STRING$(yfu - yf, 205) + "»";
FOR i = xf + 1 TO xfu
LOCATE i, yf: PRINT "°" + SPACE$(yfu - yf) + "°";
NEXT
LOCATE xfu + 1, yf: PRINT "È" + STRING$(yfu - yf, 205) + "¼";
END SUB


SUB NPkoef (k, L, r, q, h, y1, z1, v1, w1)
k = h * y1: L = h * z1
r = h * v1: q = h * w1
END SUB
```

```
SUB NPxyzvw (nk, x, x0, y, y0, z, z0, v, v0, w, w0, h, h0, k, L, r, q)
IF nk = 1 THEN
x = x0: y = y0: z = z0
v = v0: w = w0: h = h0
GOTO fund:
END IF

IF nk = 2 OR nk = 3 THEN
x = x0 + (.5 * h): y = y0 + (.5 * k)
z = z0 + (.5 * L): v = v0 + (.5 * r)
w = w0 + (.5 * q)
GOTO fund:
END IF

IF nk = 4 THEN
x = x0 + h: y = y0 + k: z = z0 + L
v = v0 + r: w = w0 + q
END IF
fund:
END SUB

SUB y1z1v1w1 (x, y, z, v, w, y1, z1, v1, w1, koef, ys, yy, wind)
yy = y * SQR(1 + z ^ 2)
IF yy > 256! THEN
y1 = -1 * koef * ((289.08 - .006328 * v) / 289.08) ^ 4.4 * (yy - 240) / (3 * yy)
ELSE
y1 = -1 * koef * ((289.08 - .006328 * v) / 289.08) ^ 4.4 * .0001212 * yy ^ 2 / yy
END IF
z1 = -9.80665 / y ^ 2
v1 = z
w1 = 1 / y
END SUB
```

3.5 PC Program Rameco.Bas, Non Standard Atmosphere

The PC program Rameco.Bas is used to find the horizontal range of a projectile and other elements of the trajectory of a projectile fired with known speed and angle when the ballistics coefficient of the projectile is also known.

The PC program Rameco.Bas is similar to the PC program Rangec. Bas, but can be used as well for the non-standard atmosphere and in presence of wind. It is a god substitute for the program Rangec.Bas, but we need to input the parameters of the standard atmosphere any time we execute the program.

Use of Rameco.Bas

Exercise 3.10 Cannon 76 mm, Corrections

In example 2.18, we found that the departure angle of a 76.2mm projectile needed to zero the cannon at the horizontal range 5060 meters is 10.00 degree.

The projectile speed is 588m/s. The ballistics coefficient is 0.7045.

Find the "correction" in horizontal range that result from a small change of each of the characteristics of the atmospheric air and bullet from the corresponding accepted standard value:

(a) Change in temperature of air, +10 degree.
(b) Change in temperature of the propellant temperature, +10 degree.
(c) Change in atmospheric pressure, +10 mm.
(d) Change in departure speed of projectile, +1% of the speed.
(e) Change in range wind, +10m/s.
(f) Change in cross wind, +10m/s).

Solution

To find one of the corrections we input the value of the parameter adding the change to the standard value.

For example, to find the correction in range that results from a change of (10mm) in the pressure of air, we input the value that results by adding (10) to the standard value of air pressure (750 degrees), i.e. we input 760 as an initial value of the pressure.

(a) **Input**: Departure speed, 588; departure angle, 10; temperature of air, 25; temperature of propellant, 15; atmospheric pressure, 750; Pressure of water vapor, 6.35; projectile mass, 1 (when mass is not known we input 1); change in projectile mass, 0; range wind, 0, cross wind, 0; ballistics coefficient, 0.7045.

   Output: Horizontal range is 5155m. The correction in range is

$$\Delta x = 5155 - 5060 = 95m.$$

In the same way, we find the other corrections.

(b) 59m; (c) -27m; (d) 47m (input initial speed, 593.88); (e) 126m; (f) 64.85m.

PC Program Rameco.Bas
Non Standard Atmosphere, Wind is Present

'FIND: Range, and other elements of trajectory.
'GIVEN: Departure Angle, Initial speed, BC
'_____

'DATA
'Input: Initial y-coordinate = 0, departure speed = 588; departure Angle = 10;
           'Temperature of Air = 25; temperature of propellant = 25;
           'Pressure = 740; Pressure of Water vapor = 6.35; Projectile mass = 1;
           'Change in Projectile mass = 0;
           'range wind, -10; cross wind, 5; integration step, 0.5.
           'BC = 0.7045
'Results: Range = 5118m, Time of Flight = 15.43s,
      'Terminal Speed = 262m/s, Terminal Angle= -16.49 Degree
      'Cross wind deflection = 31.4m; vertex (2920, 313)
'_____

'Functions & Subs.
DECLARE SUB y1z1v1w1 (x, y, z, v, w, y1, z1, v1, w1, koef, pa1, wind, ys,
      yy, pa, ta1)
DECLARE SUB InfHyres (x0, y0, z0, v0, w0, A, h0, ta, pa, ea, m, dm, tp, ta1,
      pa1, xx1, voo, vo1, wind, koef, cw, vv)
DECLARE SUB NPxyzvw (nk, x, x0, y, y0, z, z0, v, v0, w, w0, h, h0, k, L, r, q)
DECLARE SUB NPkoef (k, L, r, q, h, y1, z1, v1, w1)
DECLARE SUB menu (cog, cof, xf, yf, xfu, yfu, t$)
DECLARE SUB c (koef, m, dm, BC)

'Variables
DIM m(4, 4), v(4)
rendi = 4
cog = 7: cof = 0

'Zgjidhja
CLS
fillimi:
menu cog, cof, 3, 10, 21, 70, "INITIAL DATA"
InfHyres x0, y0, z0, v0, w0, A, h0, ta, pa, ea, m, dm, tp, ta1, pa1, xx1, voo, vo1,
      wind, koef, cw, vv
c koef, m, dm, BC
f:

```
FOR nk = 1 TO rendi
NPxyzvw nk, x, x0, y, y0, z, z0, v, v0, w, w0, h, h0, k, L, r, q
y1z1v1w1 x, y, z, v, w, y1, z1, v1, w1, koef, pa1, wind, ys, yy, pa, ta1
NPkoef k, L, r, q, h, y1, z1, v1, w1
m(nk, 1) = k: m(nk, 2) = L
m(nk, 3) = r: m(nk, 4) = q
NEXT nk

'Calculation
FOR i = 1 TO rendi
v(i) = 1 / 6 * (m(1, i) + 2 * m(2, i) + 2 * m(3, i) + m(4, i))
NEXT i

'New Data
x0 = x0 + h: y0 = y0 + v(1): z0 = z0 + v(2)
v0 = v0 + v(3): w0 = w0 + v(4)

IF ABS(z0) < .0001 THEN
ymax = v0
xmax = x0 + wind * w0
END IF
xxx = x0 + wind * w0
IF (xxx - xx1) <= .001 THEN
xc = xxx
yc = v0
tc = w0
ac = (180 / 3.141592654#) * ATN(z0)
vc = y0 / COS(ATN(z0))
END IF

IF v0 - vv <= .01 THEN
'Display Results
menu cog, cof, 6, 20, 22, 72, "RESULTS:"
LOCATE 7, 25: PRINT "Horizontal Range [m]    = "; INT((x0 + w0 * wind) + .5)
LOCATE 8, 25: PRINT "Coresponding y-Coord [m] = "; INT((v0) * 1000 + .5) / 1000
LOCATE 9, 25: PRINT "Departure Angle [Deg.]   = "; INT((A) * 1000000 +
    .5) / 1000000
LOCATE 10, 25: PRINT "Time of Flight [s]    = "; INT((w0) * 100 + .5) / 100
LOCATE 11, 25: PRINT "Terminal Speed [m/s]  = "; INT((y0 * (1 + z0^2)^.5) + .5)
```

LOCATE 12, 25: PRINT "Terminal Angle [Deg.]   = "; INT((ATN(z0) * 180 / 3.141593) * 10000 + .5) / 10000

LOCATE 13, 25: PRINT "Cross-Wind Deflection   = "; INT((cw * (w0 - x0 / (voo * COS(A * 3.14159265# / 180)))) * 1000 + .5) / 1000

LOCATE 14, 25: PRINT "Trajectory Vertex [m]   = "; "("; INT((xmax) + .5); ","; INT((ymax) + .5); ")"

LOCATE 16, 25: PRINT "Point on Trajectory [m]   = "; "("; INT((xc) + .5); ","; INT((yc) * 1000 + .5) / 1000; ")"

LOCATE 17, 25: PRINT "Time [s]   = "; INT((tc) * 100 + .5) / 100

LOCATE 18, 25: PRINT "Corresponding Speed [m/s] = "; INT((vc) + .5)

LOCATE 19, 25: PRINT "Corresponding Angle [Deg] = "; INT((ac) * 10000 + .5) / 10000

LOCATE 20, 25: PRINT "Cross-Wind Deflection   = "; INT((cw * (tc - xc / (voo * COS(A * 3.14159265# / 180)))) * 1000 + .5) / 1000

LOCATE 22, 25: PRINT "Ballistics Coefficient = "; BC

ELSE

GOTO f:

END IF

END

SUB c (koef, m, dm, BC)

BC = koef

koef = koef * (1 - dm / m)

END SUB

SUB InfHyres (x0, y0, z0, v0, w0, A, h0, ta, pa, ea, m, dm, tp, ta1, pa1, xx1, voo, vo1, wind, koef, cw, vv)

LOCATE 5, 13: INPUT "y-coordinate of Firearm   = "; v0

LOCATE 6, 13: INPUT "Departure Speed [m/s]   = "; y0

LOCATE 7, 13: INPUT "Departure Angle [Degree]   = "; z0

LOCATE 8, 13: INPUT "Temperature of Air [C]   = "; ta

LOCATE 9, 13: INPUT "Propellant Temperature[C]   = "; tp

LOCATE 10, 13: INPUT "Atmospheric Pressure [mm] = "; pa

LOCATE 11, 13: INPUT "Pressure of Water Vapor [mm] = "; ea

LOCATE 12, 13: INPUT "Projectile Mass   = "; m

LOCATE 13, 13: INPUT "Change in Projectile mass   = "; dm

LOCATE 14, 13: INPUT "Range Wind   = "; wind

LOCATE 15, 13: INPUT "Cross Wind   = "; cw

```
LOCATE 16, 13: INPUT "Ballistics Coefficient    = "; koef
LOCATE 17, 13: INPUT "x-coordinate of a point on Trajectory = "; xx1
LOCATE 18, 13: INPUT "Integration Step,  10, 1, or 0.5  = "; h0
vv = v0: A = z0: voo = y0
ta = ta + 273.15
pa1 = ta / (1 - .3785 * ea / pa)
vo1 = (voo - .4 * voo * (dm / m) + .00125 * voo * (tp - 15))
y0 = SQR(vo1 ^ 2 + wind ^ 2 - 2 * vo1 * wind * COS(a * 3.141592654# /
    180))
y0 = y0 * COS(a * 3.141592654# / 180)
z0 = TAN(A * 3.141592654# / 180)
z0 = z0 / (1 - wind / (vo1 * COS(A * 3.141592654# / 180)))
CLS
END SUB


SUB menu (cog, cof, xf, yf, xfu, yfu, t$)
COLOR cog, cof
LOCATE xf - 1, yf: PRINT t$
LOCATE xf, yf: PRINT "É" + STRING$(yfu - yf, 205) + "»";
FOR i = xf + 1 TO xfu
LOCATE i, yf: PRINT "º" + SPACE$(yfu - yf) + "º";
NEXT
LOCATE xfu + 1, yf: PRINT "È" + STRING$(yfu - yf, 205) + "¼";
END SUB


SUB NPkoef (k, L, r, q, h, y1, z1, v1, w1)
k = h * y1: L = h * z1
r = h * v1: q = h * w1
END SUB


SUB NPxyzvw (nk, x, x0, y, y0, z, z0, v, v0, w, w0, h, h0, k, L, r, q)
IF nk = 1 THEN
x = x0: y = y0: z = z0
v = v0: w = w0: h = h0
GOTO fund:
END IF


IF nk = 2 OR nk = 3 THEN
```

```
x = x0 + (.5 * h): y = y0 + (.5 * k)
z = z0 + (.5 * L): v = v0 + (.5 * r)
w = w0 + (.5 * q)
GOTO fund:
END IF

IF nk = 4 THEN
x = x0 + h: y = y0 + k: z = z0 + L
v = v0 + r: w = w0 + q
END IF
fund:
END SUB

SUB y1z1v1w1 (x, y, z, v, w, y1, z1, v1, w1, koef, pa1, wind, ys, yy, pa, ta1)
ta1 = (289.08 / pa1) ^ .5
yy = y * SQR(1 + z ^ 2)
IF yy * ta1 > 256! THEN
y1 = -1 * koef * (pa / 750) * ((pa1 - .006328 * v) / pa1) ^ 4.4 * (ta1 *
    yy - 240) / (3 * yy)
ELSE
y1 = -1 * koef * (pa / 750) * ((pa1 - .006328 * v) / pa1) ^ 4.4 * .0001212 * yy
    ^ 2 / yy
END IF
z1 = -9.80665 / y ^ 2
v1 = z
w1 = 1 / y
END SUB
```

3.6 PC Program Rcpmet.Bas, Non Standard Atmosphere

The PC program Rcpmet.Bas is similar to Rcpoint.Bas and can be used to estimate the elements of a projectile trajectory simultaneously at any two points. There are known the departure speed, the departure angle, and the ballistics coefficient of the projectile. The projectile flight is in non-standard atmosphere and in presence of wind.

The RCPoint.Bas is used as well to find the drop of a bullet fired horizontally (departure angle equal to zero degree) at any two points on the trajectory. The firearm can be at the sea level or at any altitude over the sea.

Example 3.11 Bullet Drop

The Siacci ballistics coefficient of 7.62mm M80 bullet (see section 1.1) is 3.182. The bullet is fired horizontally with a speed of 856.50m/s. The atmosphere is not standard: The temperature at the sea level is 25 degree Celsius, the temperature of propellant 25 degree Celsius, the pressure 740mm Hg, range wind 8m/s.

Use the PC program Rcpmet.Bas.

(a) Find the drop of the bullet at the points respectively with abscissa 500m and 800m if the bullet is fired horizontally from the gun located at the sea level.

(b) In the same conditions as in (a) find the drop of the bullet at the points respectively with abscissa 500 and 800 if the bullet is fired horizontally from the gun located at a mountain 1200 meters over the sea level.

Solution

(a) **Input**: Altitude of Firearm, 0; departure speed, 856.50; departure angle, 0; temperature of air, 25; temperature of propellant, 25; pressure, 740; pressure of water vapor, 6.35, mass of projectile, 1; change in mass, 0; range wind, 5; cross wind, 5; x-coordinate of a point, 500; X-coordinate of another point, 800, BC =3.182; BC, 3.182; integration step, 0.5.

Output: At the point with abscissa 500, the drop is -2.244m; at the point with abscissa 800, the drop is -7.27.

(b) In the same way, for the altitude of fire 1200 meters, we find that at the horizontal ranges 500 and 800 meters the drop of the bullet is respectively -2.155 and - 6.77.

PC Program Rcpmet.Bas
Non Standard Atmosphere, Wind is Present

'FIND: The elements of the trajectory simultaneously at two points.
        'It finds as well the Drop of a Bullet fired with departure angle equal
        to zero.
'GIVEN: Departure Speed, departure angle, BC.
'_____

'DATA
'Input: y-coordinate of FIREARM = 0, departure Angle; 0, departure speed, 856.5
        'Temperature of Air, 25, temperature of propellant, 25;
        'Pressure = 740, Pressure of Air vapor = 6.35, Projectile mass, 1;
        'Change in Projectile mass= 0, range wind, 10; cross wind, 0; BC = 3.182
        'x-coordinate of a Point= 400; X-coordinate of a Point = 600
'Results: x-coordinate= 400, y-coordinate, -1.333
        'X-coordinate= 600, y-coordinate, -3.474
'_____

'Functions & Subs.
DECLARE SUB y1z1v1w1 (x, y, z, v, w, y1, z1, v1, w1, koef, pa1, wind, ys,
    yy, pa, ta1)
DECLARE SUB InfHyres (x0, y0, z0, v0, w0, a, h0, ta, pa, ea, m, dm, tp, ta1,
    pa1, xx1, voo, vo1, wind, koef, cw, vv, xax)
DECLARE SUB NPxyzvw (nk, x, x0, y, y0, z, z0, v, v0, w, w0, h, h0, k, L, r, q)
DECLARE SUB NPkoef (k, L, r, q, h, y1, z1, v1, w1)
DECLARE SUB menu (cog, cof, xf, yf, xfu, yfu, t$)
DECLARE SUB c (koef, m, dm, BC)

'Variables
DIM m(4, 4), v(4)
rendi = 4
cog = 7: cof = 0

'Zgjidhja
CLS
fillimi:
menu cog, cof, 3, 10, 21, 70, "INITIAL DATA"
InfHyres x0, y0, z0, v0, w0, a, h0, ta, pa, ea, m, dm, tp, ta1, pa1, xx1, voo, vo1,
    wind, koef, cw, vv, xax
c koef, m, dm, BC

```
f:
FOR nk = 1 TO rendi
NPxyzvw nk, x, x0, y, y0, z, z0, v, v0, w, w0, h, h0, k, L, r, q
y1z1v1w1 x, y, z, v, w, y1, z1, v1, w1, koef, pa1, wind, ys, yy, pa, ta1
NPkoef k, L, r, q, h, y1, z1, v1, w1
m(nk, 1) = k: m(nk, 2) = L
m(nk, 3) = r: m(nk, 4) = q
NEXT nk

'Calculation
FOR i = 1 TO rendi
v(i) = 1 / 6 * (m(1, i) + 2 * m(2, i) + 2 * m(3, i) + m(4, i))
NEXT i

'New Data
x0 = x0 + h: y0 = y0 + v(1): z0 = z0 + v(2)
v0 = v0 + v(3): w0 = w0 + v(4)

IF ABS(z0) < .00001 THEN
ymax = v0
xmax = x0 + wind * w0
END IF

xxx = x0 + wind * w0
IF (xxx - xx1) <= .001 THEN
xc = xxx
yc = v0
tc = w0
ac = 180 * ATN(z0) / 3.141592654#
vc = y0 * (1 + z0 ^ 2) ^ .5
zc = cw * (w0 - xc / (voo * COS(a)))
END IF

xxT = x0 + wind * w0
IF (xxT - xax) <= .001 THEN
tt = w0
xt = xxT
yt = v0
at = 180 * ATN(z0) / 3.141592654#
```

```
vt = y0 * (1 + z0 ^ 2) ^ .5
zt = cw * (w0 - xt / (voo * COS(a)))
END IF

ytt = yt
IF ABS(v0 - ytt) >= .1 THEN
'Display Results
menu cog, cof, 6, 20, 22, 72, "RESULTS:"
LOCATE 7, 25: PRINT "x-coordinate of Point[m]  = "; INT((xc) + .5)
LOCATE 8, 25: PRINT "Coresponding y-Coord [m]  = "; INT((yc) * 1000 +
    .5) / 1000
LOCATE 9, 25: PRINT "Departure Angle [Deg.]    = "; a
LOCATE 10, 25: PRINT "Time of Flight [s]     = "; INT((tc) * 100 + .5) / 100
LOCATE 11, 25: PRINT "Terminal Speed [m/s]    = "; INT((vc) + .5)
LOCATE 12,25:PRINT"Terminal Angle [Deg.]  ="; INT((ac)*10000+.5)/10000
LOCATE 13, 25: PRINT "Cross-Wind Deflection  = "; INT((zc) * 100 + .5) / 100
LOCATE 15, 25: PRINT "Second Point[m]        = "; "("; INT((xt) + .5); ","; 
    INT((yt) * 1000 + .5) / 1000; ")"
LOCATE 16, 25: PRINT "Time [s]               = "; INT((tt) * 100 + .5) / 100
LOCATE 17, 25: PRINT "Corresponding Speed [m/s] = "; INT((vt) + .5)
LOCATE 18, 25: PRINT "Corresponding Angle [Deg] = "; INT((at) * 10000
    + .5) / 10000
LOCATE 19, 25: PRINT "Cross-Wind Deflection  = "; INT((zt) * 100 + .5) / 100
LOCATE 20, 25: PRINT "Trajectory Vertex [m]  = "; "("; INT((xmax) + .5);
    ","; INT((ymax) * 100 + .5) / 100; ")"
LOCATE 22, 25: PRINT "Ballistics Coefficient = "; BC
ELSE
GOTO f:
END IF
END

SUB c (koef, m, dm, BC)
BC = koef
koef = koef * (1 - dm / m)
END SUB

SUB InfHyres (x0, y0, z0, v0, w0, a, h0, ta, pa, ea, m, dm, tp, ta1, pa1, xx1,
    voo, vo1, wind, koef, cw, vv, xax)
LOCATE 5, 13: INPUT "y-coordinate of FIREARM    = "; v0
```

```
LOCATE 6, 13: INPUT "Departure Speed [m/s]     = "; y0
LOCATE 7, 13: INPUT "Departure Angle [Degree]   = "; z0
LOCATE 8, 13: INPUT "Temperature of Air [C]     = "; ta
LOCATE 9, 13: INPUT "Propellant Temperature[C]  = "; tp
LOCATE 10, 13: INPUT "Atmospheric Pressure [mm]  = "; pa
LOCATE 11, 13: INPUT "Pressure of Air Vapor [mm] = "; ea
LOCATE 12, 13: INPUT "Projectile Mass         = "; m
LOCATE 13, 13: INPUT "Change in Projectile mass  = "; dm
LOCATE 14, 13: INPUT "Range Wind             = "; wind
LOCATE 15, 13: INPUT "Cross Wind             = "; cw
LOCATE 16, 13: INPUT "x-coordinate of a point on Trajectory = "; xx1
LOCATE 17, 13: INPUT "x-coordinate of a point on Trajectory = "; xax
LOCATE 18, 13: INPUT "Ballistics Coefficient      = "; koef
LOCATE 19, 13: INPUT "Integration Step,  10, 1, or 0.5 = "; h0
vv = v0: a = z0: voo = y0
ta = ta + 273.15
pa1 = ta / (1 - .3785 * ea / pa)
vo1 = (voo - .4 * voo * (dm / m) + .00125 * voo * (tp - 15))
y0 = SQR(vo1 ^ 2 + wind ^ 2 - 2 * vo1 * wind * COS(a * 3.141592654# / 180))
y0 = y0 * COS(a * 3.141592654# / 180)
z0 = TAN(a * 3.141592654# / 180)
z0 = z0 / (1 - wind / (vo1 * COS(a * 3.141592654# / 180)))
CLS
END SUB


SUB menu (cog, cof, xf, yf, xfu, yfu, t$)
COLOR cog, cof
LOCATE xf - 1, yf: PRINT t$
LOCATE xf, yf: PRINT "É" + STRING$(yfu - yf, 205) + "»";
FOR i = xf + 1 TO xfu
LOCATE i, yf: PRINT "º" + SPACE$(yfu - yf) + "º";
NEXT
LOCATE xfu + 1, yf: PRINT "È" + STRING$(yfu - yf, 205) + "¼";
END SUB


SUB NPkoef (k, L, r, q, h, y1, z1, v1, w1)
k = h * y1: L = h * z1
r = h * v1: q = h * w1
```

```
END SUB

SUB NPxyzvw (nk, x, x0, y, y0, z, z0, v, v0, w, w0, h, h0, k, L, r, q)
IF nk = 1 THEN
x = x0: y = y0: z = z0
v = v0: w = w0: h = h0
GOTO fund:
END IF

IF nk = 2 OR nk = 3 THEN
x = x0 + (.5 * h): y = y0 + (.5 * k)
z = z0 + (.5 * L): v = v0 + (.5 * r)
w = w0 + (.5 * q)
GOTO fund:
END IF

IF nk = 4 THEN
x = x0 + h: y = y0 + k: z = z0 + L
v = v0 + r: w = w0 + q
END IF
fund:
END SUB

SUB y1z1v1w1 (x, y, z, v, w, y1, z1, v1, w1, koef, pa1, wind, ys, yy, pa, ta1)
ta1 = (289.08 / pa1) ^ .5
yy = y * SQR(1 + z ^ 2)
IF yy * ta1 > 256! THEN
y1 = -1 * koef * (pa / 750) * ((pa1 - .006328 * v) / pa1) ^ 4.4 * (ta1 *
    yy - 240) / (3 * yy)
ELSE
y1 = -1 * koef * (pa / 750) * ((pa1 - .006328 * v) / pa1) ^ 4.4 * .0001212 * ta1
    ^ 2 * yy ^ 2 / yy
END IF
z1 = -9.80665 / y ^ 2
v1 = z
w1 = 1 / y
END SUB
```

3.7 PC Program Range122.Bas, Standard Atmosphere

The PC program Range122.Bas, can be used to estimate the range and the other elements of the trajectory of a 122 projectile trajectory fired by the 122mm Russian cannon Mod, 1960. The departure speed, the departure angle, and the ballistics coefficient of the projectile are known, the projectile flight is in standard atmosphere and the flight is in the presence of wind. The ballistics coefficient is a function of the departure angle.

The firearm can be at the sea level or at any altitude over the sea. The accuracy of the results obtained using Range122.Bas, when the cannon is over the sea level, decreases. The same restrictions mentioned to Angle122. Bas are valid as well for Range122.Bas (For more information read section 2.3).

Use of Range122.Bas

Exercise 3.12 Cannon 122 mm, Wind Corrections

A projectile 122mm of the Russian cannon fired at an angle 13.30 degree and speed 885m/s, impacts the ground at the horizontal range 14,800 meters

The flight is in standard atmosphere and in absence of wind, the cannon is at the sea level.

(a) If the same projectile is fired in presence of an opposite range wind of 10m/s, find the "correction" in horizontal range as result of the wind.
(b) Find the correction if the wind blows in the direction of flight with the same speed, 10m/s.

Solution

Input: Y-coordinate of Cannon, 0; departure speed, 885; departure angle, 13.30; range wind, -10, cross wind, 0.
Output: Horizontal range is 14597m. The correction in range is

$$\Delta x = 14587 - 14800 = -213m.$$

(b) In the same way, we find the other correction is +205m.

PC Program Range122.Bas
Projectile 122mm, Russian Cannon 122mm, Mod. 1960
Standard Atmosphere, Wind Present

'FIND: Range, Vertex of the Trajectory, etc.
      'Elements of Trajectory for a given abscissa x
'GIVEN: Departure Angle, Projectile Speed, BC as function of departure angle
'————————————————————————————

'Control Data
'Input Data
'y-coordinate of Cannon = 0
'Input: Departure Angle: 13.3, Initial Speed = 885
'Range wind, 10; cross wind, 10; Integration Step h0 = 0.5
'Input: x-coordinate of a point on trajectory xc = 10,000
'Results:
'Range = 15005; Error in y-coordinate = -0.11; Time, 31.43s; Terminal
    Speed = 318.5
'Terminal Angle = -25.858226; Coordinates of vertex (8801, 1256), cross wind
    deflection = 144m
'Abscissa of a Point on the trajectory = 10000,
'Corresponding value y = 1219.5: Speed = 406.7; time = 17.05; angle = -3.6639
'Cross wind deflection = 54.33m
'————————————————————————————

'Functions, Subs
DECLARE SUB y1z1v1w1 (x, y, z, v, w, y1, z1, v1, w1, koef, ys, yy, wind)
DECLARE SUB InfHyres (x0, y0, z0, v0, w0, a, koef, xc1, yT, h0, wind, cw, vo)
DECLARE SUB NPxyzvw (nk, x, x0, y, y0, z, z0, v, v0, w, w0, h, h0, k, L, r, q)
DECLARE SUB NPkoef (k, L, r, q, h, y1, z1, v1, w1)
DECLARE SUB menu (cog, cof, xf, yf, xfu, yfu, t$)
DECLARE SUB c (koef, a)

'Variables
DIM m(4, 4), v(4)
rendi = 4
cog = 7: cof = 0

'Solution
CLS
fillimi:

```
menu cog, cof, 3, 10, 21, 70, "Initial Data"
InfHyres x0, y0, z0, v0, w0, a, koef, xc1, yT, h0, wind, cw, vo
c koef, a

f:
FOR nk = 1 TO rendi
NPxyzvw nk, x, x0, y, y0, z, z0, v, v0, w, w0, h, h0, k, L, r, q
y1z1v1w1 x, y, z, v, w, y1, z1, v1, w1, koef, ys, yy, wind
NPkoef k, L, r, q, h, y1, z1, v1, w1
m(nk, 1) = k: m(nk, 2) = L
m(nk, 3) = r: m(nk, 4) = q
NEXT nk

'Calculation
FOR i = 1 TO rendi
v(i) = 1 / 6 * (m(1, i) + 2 * m(2, i) + 2 * m(3, i) + m(4, i))
NEXT i

'New Data
x0 = x0 + h: y0 = y0 + v(1): z0 = z0 + v(2)
v0 = v0 + v(3): w0 = w0 + v(4)
IF y0 >= 256 THEN
IF ABS(z0) < .00001 OR ABS(z0) <= .0001 THEN
ymax = v0
xmax = x0 + wind * w0
END IF
END IF
IF y0 < 256 THEN
IF ABS(z0) < .0001 OR ABS(z0) < .001 THEN
ymax = v0
xmax = x0 + wind * w0
END IF
END IF

xab = x0 + wind * w0
IF (xab - xc1) <= .001 THEN
xc = xab
yc = v0
```

```
tc = w0
ac = (180 / 3.141592654#) * ATN(z0)
vc = y0 / COS(ATN(z0))
END IF


IF v0 - yT <= .001 THEN
'Display Results
menu cog, cof, 2, 20, 22, 76, "RESULTS:"
LOCATE 3, 26: PRINT "Departure Angle      ="; a
LOCATE 4, 26: PRINT "Departure Speed      ="; vo
LOCATE 6, 26: PRINT "Horizontal Range [m] = "; INT((x0 + wind * w0) + .5)
LOCATE 7, 26: PRINT "Error in y-coord [m] = "; INT((v0 - yT) * 1000 + .5) / 1000
LOCATE 8, 26: PRINT "Time of Flight [s]   = "; INT((w0) * 100 + .5) / 100
LOCATE 9, 26: PRINT "Terminal Speed [m/s]  = "; INT((y0 * (1 + z0 ^ 2) ^
    .5) * 100 + .5) / 100
LOCATE 10, 26: PRINT "Terminal Angle [Deg.] = "; ATN(z0) * 180 / 3.141593
LOCATE 11, 26: PRINT "Trajectory Vertex (xm, ym)   = "; "("; INT((xmax)
    * 100 + .5) / 100; ","; INT((ymax) * 100 + .5) / 100; ")"
LOCATE 12, 26: PRINT "Cross-Wind Deflection      = "; INT((cw * (w0 - x0
    / (vo * COS(a * 3.14159265# / 180)))) * 100 + .5) / 100
LOCATE 14, 23: PRINT "Abscissa of a point on trajectory:  X = "; INT((xc) + .5)
LOCATE 15, 23: PRINT "Corresponding ordinate of x:      Y = "; INT((yc)
    * 1000 + .5) / 1000
LOCATE 16, 23: PRINT "Corresponding Time:     T = "; INT((tc) * 100 + .5) / 100
LOCATE 17, 23: PRINT "Corresponding Speed:     V = "; INT((vc) * 100 + .5) / 100
LOCATE 18, 23: PRINT "Corresponding Angle:         A = "; ac
LOCATE 19, 23: PRINT "Cross-Wind Deflection         D = "; INT((cw *
    (tc - xc1 / (vo * COS(a * 3.14159265# / 180)))) * 100 + .5) / 100
LOCATE 21, 24: PRINT "Ballistics Coefficient = "; koef
ELSE
GOTO f:
END IF
END


SUB c (koef, a)
IF a >= 0 AND a <= 1.38333333# THEN koef = (1.82051 - 210.432 * (a *
    3.141592654# / 180) + 10066.5 * (a * 3.141592654# / 180) ^ 2 - 164950
    * (a * 3.141592654# / 180) ^ 3)'3200
```

```
IF a > 1.38333333# AND a <= 2.5666667# THEN koef = (.490432 - 14.0538 * (a
    * 3.141592654# / 180) + 298.238 * (a * 3.141592654# / 180) ^ 2 - 2328.38
    * (a * 3.141592654# / 180) ^ 3) '5400
IF a > 2.5666667# AND a <= 4.8833333# THEN koef = (.352739 - 4.0664 * (a
    * 3.141592654# / 180) + 49.8947 * (a * 3.141592654# / 180) ^ 2 - 207.518
    * (a * 3.141592654# / 180) ^ 3) '8600
IF a > 4.8833333# AND a < 8.0066667# THEN koef = (.304165 - 1.65929 * (a
    * 3.141592654# / 180) + 13.5957 * (a * 3.141592654# / 180) ^ 2 - 35.3659
    * (a * 3.141592654# / 180) ^ 3) '11600
IF a > 8.0066667# AND a <= 12.9166667# THEN koef = (.324563 - 1.5842 * (a
    * 3.141592654# / 180) + 9.56496 * (a * 3.141592654# / 180) ^ 2 - 17.802
    * (a * 3.141592654# / 180) ^ 3) '14600
IF a > 12.9166667# AND a <= 18.3166667# THEN koef = (.158193 + .919854 *
    (a * 3.141592654# / 180) - 3.03889 * (a * 3.141592654# / 180) ^ 2 + 3.35924
    * (a * 3.141592654# / 180) ^ 3) '17200
IF a > 18.3166667# AND a <= 45 THEN koef = (.251064 - .0064118# *
    (a * 3.141592654# / 180) + .0262451# * (a * 3.141592654# / 180) ^
    2 - .0064443# * (a * 3.141592654# / 180) ^ 3) '23800
END SUB


SUB InfHyres (x0, y0, z0, v0, w0, a, koef, xc1, yT, h0, wind, cw, vo)
LOCATE 5, 13: INPUT "y-coordinate of Cannon:          "; v0
LOCATE 6, 13: INPUT "Departure SPEED [m/s]:           "; y0
LOCATE 7, 13: INPUT "Departure ANGLE [Degree]:        "; z0
LOCATE 8, 13: INPUT "Range Wind:                      "; wind
LOCATE 9, 13: INPUT "Cross Wind:                      "; cw
LOCATE 11, 13: INPUT "X coordinate of a point on Trajectory [m]  = "; xc1
LOCATE 12, 13: INPUT "Integration Step: 10; 1; 0.5;    "; h0
yT = v0
vo = y0
a = z0
y0 = SQR(vo ^ 2 + wind ^ 2 - 2 * vo* wind * COS(a * 3.141592654# / 180))
y0 = y0 * COS(a * 3.141592654# / 180)
z0 = TAN(a * 3.141592654# / 180)
z0 = z0 / (1 - wind / (vo * COS(a * 3.141592654# / 180)))
CLS
END SUB
```

```
SUB menu (cog, cof, xf, yf, xfu, yfu, t$)
COLOR cog, cof
LOCATE xf - 1, yf: PRINT t$
LOCATE xf, yf: PRINT "É" + STRING$(yfu - yf, 205) + "»";
FOR i = xf + 1 TO xfu
LOCATE i, yf: PRINT "°" + SPACE$(yfu - yf) + "°";
NEXT
LOCATE xfu + 1, yf: PRINT "È" + STRING$(yfu - yf, 205) + "¼";
END SUB


SUB NPkoef (k, L, r, q, h, y1, z1, v1, w1)
k = h * y1: L = h * z1
r = h * v1: q = h * w1
END SUB


SUB NPxyzvw (nk, x, x0, y, y0, z, z0, v, v0, w, w0, h, h0, k, L, r, q)
IF nk = 1 THEN
x = x0: y = y0: z = z0
v = v0: w = w0: h = h0
GOTO fund:
END IF


IF nk = 2 OR nk = 3 THEN
x = x0 + (.5 * h): y = y0 + (.5 * k)
z = z0 + (.5 * L): v = v0 + (.5 * r)
w = w0 + (.5 * q)
GOTO fund:
END IF


IF nk = 4 THEN
x = x0 + h: y = y0 + k: z = z0 + L
v = v0 + r: w = w0 + q
END IF
fund:
END SUB
```

```
SUB y1z1v1w1 (x, y, z, v, w, y1, z1, v1, w1, koef, ys, yy, wind)
yy = y * SQR(1 + z ^ 2)
IF yy > 256! THEN
y1 = -1 * koef * ((289.08 - .006328 * v) / 289.08) ^ 4.4 * (yy - 240) / (3 * yy)
ELSE
y1 = -1 * koef * ((289.08 - .006328 * v) / 289.08) ^ 4.4 * .0001212 * yy ^
    2 / yy
END IF
z1 = -9.80665 / y ^ 2
v1 = z
w1 = 1 / y
END SUB
```

3.8 PC Program Rangmet.Bas, Non Standard Atmosphere

The PC program Rangmet.Bas, can be used to estimate the elements of a 122 projectile trajectory fired by the 122mm Russian cannon Mod, 1960. The departure speed, the departure angle, and the ballistics coefficient of the projectile are known and the projectile flight is in non-standard atmosphere and in presence of wind. The ballistics coefficient is function of the departure angle.

The Cannon can be at the sea level or at any altitude over the sea.

The PC program Rangmet.Bas can be used instead of the PC programs Range122.Bas considering that the projectile flight is in standard atmosphere, and that the projectile characteristics are standard.

The restrictions that are mentioned for Angmet.Bas in section 2.5 are valid as well for Rangmet.Bas.

Use of Rangmet.Bas

Exercise 3.13 Cannon 122 mm, Departure Angle Corrections

A projectile 122mm of the Russian cannon fired at an angle 13.30 degree and speed 885m/s, impacts the ground at the horizontal range 14,800 meters

The flight is in standard atmosphere and in absence of wind, and the cannon is at the sea level.

(a) The same projectile is fired in a non-standard atmosphere, temperature at the sea level +25 degree, pressure 760mm Hg, and in the presence of range wind of 10m/s that blows in direction of projectile flight. Find the "correction" in horizontal range that results from the deviation of each of those factors from the corresponding standard value (+15 degree, 750mm Hg, no range wind). Find as well the correction in horizontal range that results from the increase 0f +10 degree in the propellant temperature.

(b) Find the total change in range that corresponds to the simultaneous change of all parameters that are given in (a).

(c) Find the change in the horizontal range that corresponds to a change of -0.06 degree in the departure angle. Find as well the correction in the departure angle that is needed to set the departure angle in order to hit the target at 10000 meters in the given non-standard atmosphere.

Solution

Input: Y-coordinate of cannon, 0; departure speed, 885; departure angle, 13.30; temperature, +25; propellant temperature, 15; pressure, 750; pressure of water vapor, 6.35; mass of projectile, 1; change in range, 0; range wind,0; cross wind, 0.
Output: Horizontal range is 15029m. The correction in range is

$$\Delta x = 15029 - 14800 = +229m.$$

In the same way, we find the following corrections:
To a change of +10mm Hg, it corresponds a change of -103m in horizontal range. To a change of +10 degree in the propellant temperature, it corresponds a change of 219m in the horizontal range. To the change +10m /s in the range-wind, it corresponds a change of 216m in the horizontal range.

(b) The total change in range is 561m. (Using the Rangmet.Bas, we find a total change in range of 562m). Thus, the point of impact of the projectile fired at the angle 13.3 would be 15361 meters from the cannon.

   Since the target located at 14,800 meters will be over passed by 561 meters we need to reduce the departure angle to hit the target at 14800 meters.

(c) Using again the Rangemet.Bas, Inputting the value 13.24 of the departure angle, we find a change of (-30) meters in the horizontal range. By interpolation, we find that to correct the range the departure angle needs to be decreased by

$$\Delta \alpha = 0.06 \frac{562}{-30} = -1.124°.$$

Thus, the departure angle needed to hit the target at the horizontal range 14800 meters is 12.176 degree.

If we input the above value of the departure angle into the Rangmet. bas, we will find that the horizontal range will be 14719 meters. Thus, the projectile will miss the target and will hit the ground 81 meters in front of the target. Therefore, we have to make another correction of +0.162 degree. The new "corrected" angle is 12.338.

Using again the PC program Rangmet.Bas, we find that the range is 14813. The projectile almost will hit the target.

Repeating again the same correction procedure, we find that the departure angle is 12.314. The correction in the departure angle is (-0.986 degree).

Using the PC program Angmet.Bas we can see that the value obtained above (12.314) is the right value needed to set up the departure angle to hit the target.

Remark

The best and the most efficient way to estimate the departure angle is the PC program Angmet.Bas.

The step-by-step correction procedure presented in the above example is due to the fact that we calculated the correction of departure angle (-0.986 degree) which corresponds to the total change of 561 meters in the horizontal range, considering that there are no changes in the other parameters (temperature, pressure, wind, etc.). We assume as well that a change in a parameter does not cause changes in the other parameters.

In fact, that is not true, since, for example, a change in pressure changes the temperature as well, and so, that change in temperature causes an extra change in the horizontal range, and consequently in the departure angle. On the other hand, a change in the temperature influences the speed of sound, which, as a result, influences the drag function, and so introduces an extra change in the horizontal range.

PC Program Rangmet.Bas
Cannon 122mm, Mod. 1960. Non Standard Atmosphere

```
'FIND:    Range, and other Elements of the Trajectory, etc.
'GIVEN: Departure Speed, Departure Angle, Ballistics Coefficient
'_____

'DATA
'Input:  Initial y-coordinate =0, departure Angle; 13.3, departure speed, 885
        'Temperature of Air, 15, temperature of propellant, 15;
        'Pressure = 750, Pressure of Air vapor = 6.35, Projectile mass, 1;
        'Change in Projectile mass = 0.
'

'Results: Range = 15016, Error in y-coordinate, 0.146, Time of Flight = 31.43s,
        'Terminal Speed = 318m/s, Terminal Angle = -25.8587 Degree
        'Cross wind deflection, 144m; vertex (8801, 1256), BC = 0.2410
'_____

'Functions & Subs.
DECLARE SUB y1z1v1w1 (x, y, z, v, w, y1, z1, v1, w1, koef, pa1, wind, ys,
    yy, pa, ta1)
DECLARE SUB InfHyres (x0, y0, z0, v0, w0, a, h0, ta, pa, ea, m, dm, tp, ta1,
    pa1, xx1, voo, vo1, wind, koef, cw, vv)
DECLARE SUB NPxyzvw (nk, x, x0, y, y0, z, z0, v, v0, w, w0, h, h0, k, L, r, q)
DECLARE SUB NPkoef (k, L, r, q, h, y1, z1, v1, w1)
DECLARE SUB menu (cog, cof, xf, yf, xfu, yfu, t$)
DECLARE SUB c (koef, m, dm, BC, a)

'Variables
DIM m(4, 4), v(4)
rendi = 4
cog = 7: cof = 0

'Zgjidhja
CLS
fillimi:
menu cog, cof, 3, 10, 21, 70, "INITIAL DATA"
InfHyres x0, y0, z0, v0, w0, a, h0, ta, pa, ea, m, dm, tp, ta1, pa1, xx1, voo, vo1,
    wind, koef, cw, vv
c koef, m, dm, BC, a

f:
```

```
FOR nk = 1 TO rendi
NPxyzvw nk, x, x0, y, y0, z, z0, v, v0, w, w0, h, h0, k, L, r, q
y1z1v1w1 x, y, z, v, w, y1, z1, v1, w1, koef, pa1, wind, ys, yy, pa, ta1
NPkoef k, L, r, q, h, y1, z1, v1, w1
m(nk, 1) = k: m(nk, 2) = L
m(nk, 3) = r: m(nk, 4) = q
NEXT nk

'Calculation
FOR i = 1 TO rendi
v(i) = 1 / 6 * (m(1, i) + 2 * m(2, i) + 2 * m(3, i) + m(4, i))
NEXT i

'New Data
x0 = x0 + h: y0 = y0 + v(1): z0 = z0 + v(2)
v0 = v0 + v(3): w0 = w0 + v(4)

IF ABS(z0) < .0001 THEN
ymax = v0
xmax = x0 + wind * w0
END IF

xxc = x0 + wind * w0
IF (xxc - xx1) <= .001 THEN
xc = xxc
yc = v0
tc = w0
ac = (180 / 3.141592654#) * ATN(z0)
vc = y0 / COS(ATN(z0))
END IF

IF v0 - vv <= .01 THEN
'Display Resultst
menu cog, cof, 6, 20, 22, 72, "RESULTS:"
LOCATE 7, 25: PRINT "Horizontal Range [m]    = "; INT((x0 + w0 * wind) + .5)
LOCATE 8, 25: PRINT "Coresponding y-Coord [m] = "; INT((v0) * 1000 + .5) / 1000
LOCATE 9, 25: PRINT "Departure Angle [Deg.]    = "; INT((a) * 1000000 +
    .5) / 1000000
LOCATE 10, 25: PRINT "Time of Flight [s]    = "; INT((w0) * 100 + .5) / 100
LOCATE 11, 25: PRINT "Terminal Speed [m/s]   = "; INT((y0 * (1 + z0 ^ 2) ^ .5) + .5)
```

```
LOCATE 12, 25: PRINT "Terminal Angle [Deg.]    = "; INT((ATN(z0) * 180
    / 3.141593) * 10000 + .5) / 10000
LOCATE 13, 25: PRINT "Cross-Wind Deflection    = "; INT((cw * (w0 - x0 /
    (voo * COS(a * 3.14159265# / 180)))) * 1000 + .5) / 1000
LOCATE 14, 25: PRINT "Trajectory Vertex [m]    = "; "("; INT((xmax) + .5);
    ","; INT((ymax) + .5); ")"
LOCATE 16, 25: PRINT "Point on Trajectory [m]    = "; "("; INT((xc) + .5);
    ","; INT((yc) * 1000 + .5) / 1000; ")"
LOCATE 17, 25: PRINT "Time [s]                 = "; INT((tc) * 100 + .5) / 100
LOCATE 18, 25: PRINT "Corresponding Speed [m/s] = "; INT((vc) + .5)
LOCATE 19, 25: PRINT "Corresponding Angle [Deg] = "; INT((ac) * 10000
    + .5) / 10000
LOCATE 20, 25: PRINT "Cross-Wind Deflection    = "; INT((cw * (tc - xc /
    (voo * COS(a * 3.14159265# / 180)))) * 1000 + .5) / 1000
LOCATE 22, 25: PRINT "Ballistics Coefficient = "; BC
ELSE
GOTO f:
END IF
END


SUB c (koef, m, dm, BC, a)
IF a >= 0 AND a <= 1.38333333# THEN koef = (1 - dm / m) * (1.82051 - 210.432
    * (a * 3.141592654# / 180) + 10066.5 * (a * 3.141592654# / 180) ^
    2 - 164950 * (a * 3.141592654# / 180) ^ 3)'3200
IF a > 1.38333333# AND a <= 2.5666667# THEN koef = (1 - dm / m)
    * (.490432 - 14.0538 * (a * 3.141592654# / 180) + 298.238 * (a *
    3.141592654# / 180) ^ 2 - 2328.38 * (a * 3.141592654# / 180) ^ 3) '5400
IF a > 2.5666667# AND a <= 4.8833333# THEN koef = (1 - dm / m)
    * (.352739 - 4.0664 * (a * 3.141592654# / 180) + 49.8947 * (a *
    3.141592654# / 180) ^ 2 - 207.518 * (a * 3.141592654# / 180) ^ 3) '8600
IF a > 4.8833333# AND a < 8.0066667# THEN koef = (1 - dm / m) *
    (.304165 - 1.65929 * (a * 3.141592654# / 180) + 13.5957 * (a *
    3.141592654# / 180) ^ 2 - 35.3659 * (a * 3.141592654# / 180) ^ 3) '11600
IF a > 8.0066667# AND a <= 12.9166667# THEN koef = (1 - dm / m)
    * (.324563 - 1.5842 * (a * 3.141592654# / 180) + 9.56496 * (a *
    3.141592654# / 180) ^ 2 - 17.802 * (a * 3.141592654# / 180) ^ 3)'14600
IF a > 12.9166667# AND a <= 18.3166667# THEN koef = (1 - dm / m)
    * (.158193 + .919854 * (a * 3.141592654# / 180) - 3.03889 * (a *
    3.141592654# / 180) ^ 2 + 3.35924 * (a * 3.141592654# / 180) ^ 3) '17200
```

```
IF a > 18.3166667# AND a <= 45 THEN koef = (1 - dm / m) *
    (.251064 - .0064118# * (a * 3.141592654# / 180) + .0262451# * (a *
    3.141592654# / 180) ^ 2 - .0064443# * (a * 3.141592654# / 180) ^ 3) '23800
BC = koef / (1 - dm / m)
END SUB


SUB InfHyres (x0, y0, z0, v0, w0, a, h0, ta, pa, ea, m, dm, tp, ta1, pa1, xx1,
    voo, vo1, wind, koef, cw, vv)
LOCATE 5, 13: INPUT "y-coordinate of Firearm         = "; v0
LOCATE 6, 13: INPUT "Departure Speed [m/s]           = "; y0
LOCATE 7, 13: INPUT "Departure Angle [Degree]        = "; z0
LOCATE 8, 13: INPUT "Temperature of Air [C]          = "; ta
LOCATE 9, 13: INPUT "Propellant Temperature[C]       = "; tp
LOCATE 10, 13: INPUT "Atmospheric Pressure [mm]      = "; pa
LOCATE 11, 13: INPUT "Pressure of Water Vapor [mm]   = "; ea
LOCATE 12, 13: INPUT "Projectile Mass                = "; m
LOCATE 13, 13: INPUT "Change in Projectile mass      = "; dm
LOCATE 14, 13: INPUT "Range Wind            = "; wind
LOCATE 15, 13: INPUT "Cross Wind            = "; cw
LOCATE 16, 13: INPUT "x-coordinate of a point on Trajectory = "; xx1
LOCATE 17, 13: INPUT "Integration Step,  10, 1, or 0.5 = "; h0
vv = v0: a = z0: voo = y0
ta = ta + 273.15
pa1 = ta / (1 - .3785 * ea / pa)
vo1 = (voo - .4 * voo * (dm / m) + .00125 * voo * (tp - 15))
y0 = SQR(vo1 ^ 2 + wind ^ 2 - 2 * vo1 * wind * COS(a * 3.141592654# / 180))
y0 = y0 * COS(a * 3.141592654# / 180)
z0 = TAN(a * 3.141592654# / 180)
z0 = z0 / (1 - wind / (vo1 * COS(a * 3.141592654# / 180)))
CLS
END SUB


SUB menu (cog, cof, xf, yf, xfu, yfu, t$)
COLOR cog, cof
LOCATE xf - 1, yf: PRINT t$
LOCATE xf, yf: PRINT "É" + STRING$(yfu - yf, 205) + "»";
FOR i = xf + 1 TO xfu
LOCATE i, yf: PRINT "º" + SPACE$(yfu - yf) + "º";
NEXT
```

```
LOCATE xfu + 1, yf: PRINT "È" + STRING$(yfu - yf, 205) + "¼";
END SUB

SUB NPkoef (k, L, r, q, h, y1, z1, v1, w1)
k = h * y1: L = h * z1
r = h * v1: q = h * w1
END SUB

SUB NPxyzvw (nk, x, x0, y, y0, z, z0, v, v0, w, w0, h, h0, k, L, r, q)
IF nk = 1 THEN
x = x0: y = y0: z = z0
v = v0: w = w0: h = h0
GOTO fund:
END IF

IF nk = 2 OR nk = 3 THEN
x = x0 + (.5 * h): y = y0 + (.5 * k)
z = z0 + (.5 * L): v = v0 + (.5 * r)
w = w0 + (.5 * q)
GOTO fund:
END IF

IF nk = 4 THEN
x = x0 + h: y = y0 + k: z = z0 + L
v = v0 + r: w = w0 + q
END IF
fund:
END SUB

SUB y1z1v1w1 (x, y, z, v, w, y1, z1, v1, w1, koef, pa1, wind, ys, yy, pa, ta1)
ta1 = (pa1 / pa1) ^ .5yy = y * SQR(1 + z ^ 2)
IF yy * ta1 > 256! THEN
y1 = -1 * koef * (pa / 750) * ((pa1 - .006328 * v) / pa1) ^ 4.4 * (ta1 *
    yy - 240) / (3 * yy)
ELSE
y1 = -1 * koef * (pa / 750) * ((pa1 - .006328 * v) / pa1) ^ 4.4 * .0001212 *
    ta1^2 * yy ^ 2 / yy
END IF
z1 = -9.80665 / y ^ 2
v1 = z
w1 = 1 / y
END SUB
```

3.9 PC Program Rangem30.Bas, Non Standard Atmosphere

The PC program RangeM30.Bas can be employed to compute the horizontal range and other elements of the trajectory of a 0.30 Ball M2 bullet fired with speed 853.44m/s (2800fps), in standard atmosphere. The wind is present.

The gun and the target are at the same altitude over the sea level. The ballistics coefficient is a function of the projectile speed.

Using Rangem30.bas we can compute as well the elements of the trajectory at a point with a given abscissa.

Use of PC program Rangem30.Bas

Example 3.14

A bullet 0.30 Ball M2 is fired with speed 853.44m/s at an angle 0.2755 degree in standard atmosphere. The gun and the target are at the sea level.

(a) Find the horizontal range and the other elements of the trajectory at the terminal point. Find as well the elements of the trajectory at the point with abscissa 300 meters.
(b) Find the drop of the projectile at 300 meters and 500 meters
(c) What is the drop of the bullet at 300 meters and 500 meters if the bullet is fired horizontally, i.e. departure angle is zero degree?
(d) Find the range and the cross deviation if there is a range wind of 6m/s and a cross wind 4m/s .

Solution

(a) **Input**: Altitude of gun, 0; departure speed, 853.44; departure angle, 0.2755; range wind, 0; cross wind, 0; x-coordinate of a trajectory point, 300; integration step, 0.5.

Output: Horizontal range, 500m; error in y-coordinate, -0.001m; time, 0.76s; terminal speed, 508m/s; terminal angle, -0.3897 degree; trajectory vertex is located at the point with coordinates (275.5m, 0.71m).

For the point with abscissa 300m the corresponding y-coordinate is 0.702m; time, 0.42; speed, 635.5m; angle, -0.03852.

(b) The drop of the bullet at the 500 meters is

$$\bar{y}_5 = 500 \cdot \tan(0.2755) = 2.404m$$

The drop of the bullet at 300 meters is

$$\bar{y}_3 = 300 \cdot \tan(.2755) - 0.702 = 0.741m$$

(c) If the bullet is fired horizontally the drop of the bullet at 300m and 500 m, is equal to the drop of the bullet calculated in (b) when the bullet is fired at an angle of 0.2755 degree, i.e. respectively 0.741m and 2.404.

   The above results are obvious based on the property expressed by the equation (1.5.1), section 5.1.

(d) The horizontal range is 501.57m; terminal angle, 0.3918 degree; cross deflection is 0.714m.

Note. The deviation in the vertical direction caused by the wind is

$$\Delta y = 1.57 \cdot \tan(-0.3918) = 0.01m = 1cm$$

As we see the deviation in the vertical direction caused by the range wind is insignificant.

PC Program Rangem30.Bas
Standard Atmosphere, Wind Present
Bullet 0.30 Ball M2 150 grain

```
'FIND: Range, Vertex of the Trajectory, etc.
        'Elements of Trajectory for a given abscisa x
'GIVEN: Launching Angle, Ballistics Coefficient, Projectile Speed


'_____

'Control Data

'Input Data
'Input: Altitude of Gun
'Input: Departure Angle: 0.25522, Departure Speed = 883.44,
'Input: Initial Time t0 = 0, Integration Step h0 = (10, 1 or 0.5, 0.1), 0.5
'Input: x-coordinate of a point on trajectory 0.
'Results
'Range = 500; Error in y-coordinate = 0; Time= 0.73s, Terminal Speed = 530.41
'Terminal Angle = -0.359; Coordinates of vertex (276, 066)
'_____

'Functions, Subs
DECLARE SUB y1z1v1w1 (x, y, z, v, w, y1, z1, v1, w1, koef, ys, yy, wind)
DECLARE SUB InfHyres (x0, y0, z0, v0, w0, a, koef, xc1, yT, h0, wind, cw, vo)
DECLARE SUB NPxyzvw (nk, x, x0, y, y0, z, z0, v, v0, w, w0, h, h0, k, L, r, q)
DECLARE SUB NPkoef (k, L, r, q, h, y1, z1, v1, w1)
DECLARE SUB menu (cog, cof, xf, yf, xfu, yfu, t$)
DECLARE SUB c (koef, y0)

'Variables
DIM m(4, 4), v(4)
rendi = 4
cog = 7: cof = 0

'Solution
CLS

fillimi:
menu cog, cof, 3, 10, 21, 70, "Initial Data"
```

```
InfHyres x0, y0, z0, v0, w0, a, koef, xc1, yT, h0, wind, cw, vo
c koef, y0

f:
FOR nk = 1 TO rendi
NPxyzvw nk, x, x0, y, y0, z, z0, v, v0, w, w0, h, h0, k, L, r, q
y1z1v1w1 x, y, z, v, w, y1, z1, v1, w1, koef, ys, yy, wind
NPkoef k, L, r, q, h, y1, z1, v1, w1
m(nk, 1) = k: m(nk, 2) = L
m(nk, 3) = r: m(nk, 4) = q
NEXT nk

'Calculation
FOR i = 1 TO rendi
v(i) = 1 / 6 * (m(1, i) + 2 * m(2, i) + 2 * m(3, i) + m(4, i))
NEXT i

'New Data
x0 = x0 + h: y0 = y0 + v(1): z0 = z0 + v(2)
v0 = v0 + v(3): w0 = w0 + v(4)
IF y0 >= 256 THEN
IF ABS(z0) < .00001 OR ABS(z0) <= .0001 THEN
ymax = v0
xmax = x0
END IF
END IF
IF y0 < 256 THEN
IF ABS(z0) < .0001 OR ABS(z0) < .001 THEN
ymax = v0
xmax = x0
END IF
END IF

IF ABS(x0 - xc1) <= .01 THEN
xc = x0
yc = v0
tc = w0
ac = (180 / 3.141592654#) * ATN(z0)
vc = y0 / COS(ATN(z0))
END IF
```

```
IF v0 - yT <= .001 THEN
'Display Results
menu cog, cof, 2, 20, 22, 76, "RESULTS:"

LOCATE 3, 26: PRINT "Departure Angle        ="; a
LOCATE 4, 26: PRINT "Departure Speed        ="; vo
LOCATE 6, 26: PRINT "Horizontal Range  [m] ="; INT((wind * w0 + x0) *
    100 + .5) / 100
LOCATE 7, 26: PRINT "Error in y-coord [m]  ="; INT((v0 - yT) * 1000 + .5) / 1000
LOCATE 8, 26: PRINT "Time of Flight [s]    = "; INT((w0) * 100 + .5) / 100
LOCATE 9, 26: PRINT "Terminal Speed [m/s]  = "; INT((y0 * (1 + z0 ^ 2) ^
    .5) * 100 + .5) / 100
LOCATE 10, 26: PRINT "Terminal Angle [Deg.]  = "; ATN(z0) * 180 /
    3.141593
LOCATE 11, 26: PRINT "Trajectory Vertex (xm, ym)  = "; "("; INT((xmax)
    * 100 + .5) / 100; ","; INT((ymax) * 100 + .5) / 100; ")"
LOCATE 12, 26: PRINT "Cross-Wind Deflection      = "; INT((cw * (w0 - x0
    / (vo * COS(a * 3.14159265# / 180)))) * 1000 + .5) / 1000
LOCATE 14, 23: PRINT "Abscissa of a point on trajectory:  X = "; INT((wind
    * tc + xc) * 100 + .5) / 100
LOCATE 15, 23: PRINT "Corresponding ordinate of x:      Y = "; INT((yc)
    * 1000 + .5) / 1000
LOCATE 16, 23: PRINT "Corresponding Time:            T = "; INT((tc) *
    100 + .5) / 100
LOCATE 17, 23: PRINT "Corresponding Speed:          V = "; INT((vc) *
    100 + .5) / 100
LOCATE 18, 23: PRINT "Corresponding Angle:          A = "; ac
LOCATE 19, 23: PRINT "Cross-Wind Deflection         Z = "; INT((cw *
    (tc - xc / (vo * COS(a * 3.14159265# / 180)))) * 1000 + .5) / 1000
LOCATE 21, 24: PRINT "Ballistics Coefficient = "; koef
ELSE
GOTO f:
END IF
END


SUB c (koef, y0)
koef = (.913405 - 9.944000000000001D-04 * y0 + .00000062# * y0 ^ 2) *
    6.3223913#
END SUB
```

```
SUB InfHyres (x0, y0, z0, v0, w0, a, koef, xc1, yT, h0, wind, cw, vo)
LOCATE 5, 13: INPUT "Altitude of the GUN       = "; v0
LOCATE 6, 13: INPUT "Departure Speed [m/s]     = "; y0
LOCATE 7, 13: INPUT "Launching Angle [Degree]    = "; z0
LOCATE 8, 13: INPUT "Range Wind            = "; wind
LOCATE 9, 13: INPUT "Cross Wind            = "; cw
LOCATE 10, 13: INPUT "Integration Step: 10; 1; 0.5     = "; h0
LOCATE 11, 13: INPUT "x-coordinate of a Trajectory POINT = "; xc1
yT = v0
vo = y0
a = z0
y0 = SQR(vo ^ 2 + wind ^ 2 - 2 * vo * wind * COS(a * 3.141592654# / 180))
y0 = y0 * COS(a * 3.141592654# / 180)
z0 = TAN(a * 3.141592654# / 180)
z0 = z0 / (1 - wind / (vo * COS(a * 3.141592654# / 180)))
CLS
END SUB


SUB menu (cog, cof, xf, yf, xfu, yfu, t$)
COLOR cog, cof
LOCATE xf - 1, yf: PRINT t$
LOCATE xf, yf: PRINT "É" + STRING$(yfu - yf, 205) + "»";
FOR i = xf + 1 TO xfu
LOCATE i, yf: PRINT "°" + SPACE$(yfu - yf) + "°";
NEXT
LOCATE xfu + 1, yf: PRINT "È" + STRING$(yfu - yf, 205) + "¼";
END SUB


SUB NPkoef (k, L, r, q, h, y1, z1, v1, w1)
k = h * y1: L = h * z1
r = h * v1: q = h * w1
END SUB


SUB NPxyzvw (nk, x, x0, y, y0, z, z0, v, v0, w, w0, h, h0, k, L, r, q)
IF nk = 1 THEN
x = x0: y = y0: z = z0
```

```
v = v0: w = w0: h = h0
GOTO fund:
END IF
IF nk = 2 OR nk = 3 THEN
x = x0 + (.5 * h): y = y0 + (.5 * k)
z = z0 + (.5 * L): v = v0 + (.5 * r)
w = w0 + (.5 * q)
GOTO fund:
END IF

IF nk = 4 THEN
x = x0 + h: y = y0 + k: z = z0 + L
v = v0 + r: w = w0 + q
END IF
fund:
END SUB

SUB y1z1v1w1 (x, y, z, v, w, y1, z1, v1, w1, koef, ys, yy, wind)
yy = y * SQR(1 + z ^ 2)
IF yy > 256! THEN
y1 = -1 * koef * ((289.08 - .006328 * v) / 289.08) ^ 4.4 * (yy - 240) / (3 * yy)
ELSE
y1 = -1 * koef * ((289.08 - .006328 * v) / 289.08) ^ 4.4 * .0001212 * yy ^
    2 / yy
END IF
z1 = -9.80665 / y ^ 2
v1 = z
w1 = 1 / y
END SUB
```

3.10 PC Program Parach.Bas, Skydiving in Standard Atmosphere

The PC program Parach.Bas estimates the trajectory of a parachutist launched usually horizontally (launching angle zero) from an airplane, or helicopter with the initial speed equal to the speed of the airplane or helicopter at the instant the skydiver jumps from the plane. We have not considered the speed of jump in the direction of jumping.

For more information, refer to chapter 9 of Exterior Ballistics with Applications.

Use of PC Program Parach.Bas

Example 3.15

A skydiver of mass 105kg jumps from an airplane that flies horizontally with a speed 110m/s at an altitude 4000 meters over the sea level. The parachutist deploys the parachute when he/she is 2400 meters over the sea level.

The atmosphere is standard but there is a range wind of 20m/s, and a cross wind of 10m/s.

Find:

(a) The elements of the trajectory of flight at the altitude1200 meters;
(b) Find the elements of the trajectory after10 seconds.

Input Data: Altitude of launching point, 2500; altitude of parachute deployment, 1200; launching speed, 100; launching angle, 0; range wind, 10; time of interest, 8; temperature of air, 5; atmospheric pressure, 760; coefficient "b" is 0.241; mass of skydiver, 100; Integration Step, 0.01.

Results

(a) x-coordinate of skydiver, 1276m; y-coordinate of skydiver, 2400m; time of flight, 27.61s; speed of skydiver, 73m/s; falling angle,-87.63 degree; cross deflection, 155m.
(b) After 10 seconds the skydiver location is at the point with coordinates (715m, 3644m); corresponding speed, 67m/s; corresponding angle, -64.47 degree; cross deflection, 32m.

PC Program Parach.Bas
Standard Atmosphere, Wind Present

'FIND: Coordinates of Parachutist (JUMPING With Closed Parachute)
'GIVEN: Launching Angle, Drag Coefficient, Initial Speed
'_____

'Control Data

'Input Data

'Input: Jumping Altitude [m] = 4000:
'Altitude of Parachute Deployment [m] = 2800
'Launching Speed [m/s] = 100,
'Input: Launching Angle [Degree]: 0,
'Input: Time of Interest [s] = 10
'Input: Drag Coefficient = 0.241
'Input: Mass of Parachutist [kg] = 100
'Input: Range wind = 20m/s
'Input: Cross wind = 15
'Input: Integration Step h0 = 0.01

'OUTPUT
'Jumping Altitude [m] = 4000m:
'Altitude of Parachute Deployment = 2800m; Corresponding Abscissa = 1117m
'Speed of parachutist = 73m/s; Falling Angle = -85.3 degree; Time of Flight = 22s
'Cross Deflection = 164m
'At Time T = 10; Parachutist Location is (697, 3642)
'Corresponding Speed = 67; Corresponding Angle = -65.181
'Cross Deflection = 46.
'Ballistics Coefficient is = 0.0241
'_____

'Functions, Subs
DECLARE SUB y1z1v1w1 (x, y, z, v, w, y1, z1, v1, w1, koef)
DECLARE SUB InfHyres (x0, y0, z0, v0, w0, a, koef, xc1, h0, vo, mas, xx1, alt, time, cd, wind, cw, vo1)
DECLARE SUB NPxyzvw (nk, x, x0, y, y0, z, z0, v, v0, w, w0, h, h0, k, L, r, q)
DECLARE SUB NPkoef (k, L, r, q, h, y1, z1, v1, w1)
DECLARE SUB menu (cog, cof, xf, yf, xfu, yfu, t$)
DECLARE SUB c (koef)

```
'Variables
DIM m(4, 4), v(4)
rendi = 4
cog = 7: cof = 0

'Solution
CLS
fillimi:
menu cog, cof, 3, 10, 21, 70, "Initial Data"
InfHyres x0, y0, z0, v0, w0, a, koef, xc1, h0, vo, mas, xx1, alt, time, cd, wind, cw, vol
c koef

f:
FOR nk = 1 TO rendi
NPxyzvw nk, x, x0, y, y0, z, z0, v, v0, w, w0, h, h0, k, L, r, q
y1z1v1w1 x, y, z, v, w, y1, z1, v1, w1, koef
NPkoef k, L, r, q, h, y1, z1, v1, w1
m(nk, 1) = k: m(nk, 2) = L
m(nk, 3) = r: m(nk, 4) = q
NEXT nk

'Calculation
FOR i = 1 TO rendi
v(i) = 1 / 6 * (m(1, i) + 2 * m(2, i) + 2 * m(3, i) + m(4, i))
NEXT i

'New Data
x0 = x0 + h: y0 = y0 + v(1): z0 = z0 + v(2)
v0 = v0 + v(3): w0 = w0 + v(4)

IF ABS(v0 - xx1) <= 1 THEN
yc = xx1
tc = w0
xc = x0 + wind * tc
ac = 180 * ATN(z0) / 3.141592654#
vc = y0 * SQR(1 + z0 ^ 2)
END IF
```

```
IF w0 >= time AND w0 <= time + .01 THEN
tt = w0
xt = x0 + tt * wind
yt = v0
at = 180 * ATN(z0) / 3.141592654#
vt = y0 * (1 + z0 ^ 2) ^ .5
END IF


IF z0 >= -6000 AND z0 <= -5000 THEN
'Display Results
menu cog, cof, 2, 20, 22, 76, "RESULTS:"
LOCATE 4, 23: PRINT "JUMPING ALTITUDE                = "; vo
LOCATE 5, 23: PRINT "ALTITUDE OF PARACHUTE DEPLOYMENT:   y
    = "; INT((yc) + .5)
LOCATE 6, 23: PRINT "Corresponding Abscissa x:  x = "; INT((xc) + .5)
LOCATE 7, 23: PRINT "Corresponding Time:      t = "; INT((tc) * 100 + .5) / 100
LOCATE 8, 23: PRINT "Corresponding Speed:       v = "; INT((vc) + .5)
LOCATE 9, 23: PRINT "Corresponding Angle:   a = "; INT((ac) * 1000 + .5) / 1000
LOCATE 10, 23: PRINT "Cross DEFLECTION:        z = "; INT((cw * (tc
    - xc / vo1)) + .5)
LOCATE 12, 23: PRINT "AT TIME: T             : "; INT((tt) * 100 + .5) / 100
LOCATE 13, 23: PRINT "Parachutist Location   : "; "("; INT((xt) + .5); ",";
    INT((yt) + .5); ")"
LOCATE 14, 23: PRINT "Corresponding Speed : "; INT((vt) + .5)
LOCATE 15, 23: PRINT "Corresponding Angle : "; INT((at) * 1000 + .5) / 1000
LOCATE 16, 23: PRINT "Cross DEFLECTION   : "; INT((cw * (tt - xt / vo1)) + .5)
ELSE
GOTO f:
END IF
END


SUB c (koef)
koef = koef
END SUB


SUB InfHyres (x0, y0, z0, v0, w0, a, koef, xc1, h0, vo, mas, xx1, alt, time, cd,
    wind, cw, vo1)
LOCATE 5, 13: INPUT "Launching Altitude [m]       = "; v0
```

```
LOCATE 6, 13: INPUT "Altitude of Parachute Deployment = "; xx1
LOCATE 7, 13: INPUT "Launching Speed [m/s]        = "; y0
LOCATE 8, 13: INPUT "Launching Angle [Degree]     = "; z0
LOCATE 9, 13: INPUT "Range Wind Speed             = "; wind
LOCATE 10, 13: INPUT "Cross Wind Speed            = "; cw
LOCATE 11, 13: INPUT "Time of Interest            = "; time
LOCATE 12, 13: INPUT "Coefficient b               = "; cd
LOCATE 13, 13: INPUT "Mass of Parachutist         = "; mas
LOCATE 14, 13: INPUT "Integration Step: 0.01      = "; h0
vo = v0: alt = v0: vo1 = y0: a = z0: xc1 = 0: koef = cd / mas
y0 = SQR(vo1 ^ 2 + wind ^ 2 - 2 * vo1 * wind * COS(a * 3.141516954# / 180))
y0 = SQR(y0 ^ 2 + cw ^ 2)
y0 = y0 * COS(a * 3.141516954# / 180)
z0 = TAN(z0 * 3.141516954# / 180)
z0 = z0 / (1 - wind / (vo1 * COS(a * 3.141516954# / 180)))
CLS
END SUB


SUB menu (cog, cof, xf, yf, xfu, yfu, t$)
COLOR cog, cof
LOCATE xf - 1, yf: PRINT t$
LOCATE xf, yf: PRINT "É" + STRING$(yfu - yf, 205) + "»";
FOR i = xf + 1 TO xfu
LOCATE i, yf: PRINT "º" + SPACE$(yfu - yf) + "º";
NEXT
LOCATE xfu + 1, yf: PRINT "È" + STRING$(yfu - yf, 205) + "¼";
END SUB


SUB NPkoef (k, L, r, q, h, y1, z1, v1, w1)
k = h * y1: L = h * z1
r = h * v1: q = h * w1
END SUB


SUB NPxyzvw (nk, x, x0, y, y0, z, z0, v, v0, w, w0, h, h0, k, L, r, q)
IF nk = 1 THEN
x = x0: y = y0: z = z0
v = v0: w = w0: h = h0
GOTO fund:
END IF
```

```
IF nk = 2 OR nk = 3 THEN
x = x0 + (.5 * h): y = y0 + (.5 * k)
z = z0 + (.5 * L): v = v0 + (.5 * r)
w = w0 + (.5 * q)
GOTO fund:
END IF

IF nk = 4 THEN
x = x0 + h: y = y0 + k: z = z0 + L
v = v0 + r: w = w0 + q
END IF
fund:
END SUB

SUB y1z1v1w1 (x, y, z, v, w, y1, z1, v1, w1, koef)
yy = y * SQR(1 + z ^ 2)
y1 = -1 * koef * ((289.08 - .006328 * (v)) / 289.08) ^ 4.4 * (yy ^ 2) / (yy)
z1 = -9.80665 / y ^ 2
v1 = z
w1 = 1 / y
END SUB
```

3.11 PC Program Paramet.Bas, Skydiving in Non-Standard Atmosphere

The PC program is a modified version of the program Parach.Bas to be used when the skydiving is in a non-standard atmosphere.

For more information, refer to chapter 9 of Exterior Ballistics with Applications.

Use of the PC Program Paramet.Bas

Example 3.16

(Ref. Example 1, page 491, Exterior Ballistics with Applications.)

A skydiver of mass 100kg jumps from an airplane that flies horizontally with a speed 100m/s at an altitude 2500m. The parachutist deploys the parachute when he/she is $y = 1200m$ over the sea level. Find:

(a) The elements of the trajectory of flight at the altitude $y = 1200m$;
(b) Find the elements of the trajectory after $t = 8s$.

Input Data: Altitude of launching point, 2500; altitude of parachute deployment, 1200; launching speed, 100; launching angle, 0; range wind, 10; time of interest, 8; temperature of air, 5; atmospheric pressure, 760; coefficient "b", 0.241; mass of skydiver, 100; Integration Step, 0.01.

Results

(a) x-coordinate of skydiver, 891m; y-coordinate of skydiver, 1200m; time of flight, 25s; speed of skydiver, 65m/s; falling angle,-87.60 degree.
(b) After 8 seconds the skydiver location is at the point with coordinates (510m, 2270m); corresponding speed, 59m/s; corresponding angle, -57.90 degree.

PC Program Paramet.Bas
Non Standard Atmosphere, Wind Present

'FIND: Coordinates of Parachutist (JUMPING With Closed Parachute)
'GIVEN: Launching Angle, Drag Coefficient, Initial Speed
'_____

'Control Data
'Input Data
'Input: Jumping Altitude [m] = 4000:
'Altitude of Parachute Deployment [m] = 2800
'Launching Speed [m/s] = 100,

'Input: Launching Angle [Degree]: 0,
'Input: Time of Interest [s] = 10
'Input: Drag Coefficient = 0.241
'Input: Mass of Parachutist [kg] = 100
'Input: Range wind = 20m/s
'Input: Cross wind = 15
'Input: Integration Step h0 = 0.01
'OUTPUT

'Jumping Altitude [m] = 4000m:
'Altitude of Parachute Deployment = 2800m; Corresponding Abscissa = 1117m
'Speed of parachutist = 73m/s; Falling Angle = -85.3 degree; Time of Flight = 22s
'Cross Deflection = 164m
'At Time T = 10; Parachutist Location is (697, 3642)
'Corresponding Speed = 67; Corresponding Angle = -65.181
'Cross Deflection = 46.
'Ballistics Coefficient is = 0.0241
'_____

'Functions, Subs
DECLARE SUB y1z1v1w1 (x, y, z, v, w, y1, z1, v1, w1, koef, ta1, pa)
DECLARE SUB InfHyres (x0, y0, z0, v0, w0, a, koef, xc1, h0, vo, mas, xx1,
    alt, time, cd, Wind, cw, vo1, ta, ta1, pa, ea, pa1)
DECLARE SUB NPxyzvw (nk, x, x0, y, y0, z, z0, v, v0, w, w0, h, h0, k, L, r, q)
DECLARE SUB NPkoef (k, L, r, q, h, y1, z1, v1, w1)
DECLARE SUB menu (cog, cof, xf, yf, xfu, yfu, t$)
DECLARE SUB c (koef, a)

```
'Variables
DIM m(4, 4), v(4)
rendi = 4
cog = 7: cof = 0

'Solution
CLS
fillimi:
menu cog, cof, 3, 10, 21, 70, "Initial Data"
InfHyres x0, y0, z0, v0, w0, a, koef, xc1, h0, vo, mas, xx1, alt, time, cd, Wind,
    cw, vo1, ta, ta1, pa, ea, pa1
c koef, a

f:
FOR nk = 1 TO rendi
NPxyzvw nk, x, x0, y, y0, z, z0, v, v0, w, w0, h, h0, k, L, r, q
y1z1v1w1 x, y, z, v, w, y1, z1, v1, w1, koef, ta1
NPkoef k, L, r, q, h, y1, z1, v1, w1
m(nk, 1) = k: m(nk, 2) = L
m(nk, 3) = r: m(nk, 4) = q
NEXT nk

'Calculation
FOR i = 1 TO rendi
v(i) = 1 / 6 * (m(1, i) + 2 * m(2, i) + 2 * m(3, i) + m(4, i))
NEXT i

'New Data
x0 = x0 + h: y0 = y0 + v(1): z0 = z0 + v(2)
v0 = v0 + v(3): w0 = w0 + v(4)

IF ABS(v0 - xx1) <= 1 THEN
yc = xx1
tc = w0
xc = x0 + Wind * tc
ac = 180 * ATN(z0) / 3.141592654#
vc = y0 * SQR(1 + z0 ^ 2)
END IF
```

```
IF w0 >= time AND w0 <= time + .01 THEN
tt = w0
xt = x0 + tt * Wind
yt = v0
at = 180 * ATN(z0) / 3.141592654#
vt = y0 * (1 + z0 ^ 2) ^ .5
END IF


IF z0 >= -6000 AND z0 <= -5000 THEN
'Display Results
menu cog, cof, 2, 20, 22, 76, "RESULTS:"
LOCATE 4, 23: PRINT "JUMPING ALTITUDE        ="; vo
LOCATE 5, 23: PRINT "ALTITUDE OF PARACHUTE DEPLOYMENT:   y
    = "; INT((yc) + .5)
LOCATE 6, 23: PRINT "Corresponding Abscissa:    x = "; INT((xc) + .5)
LOCATE 7, 23: PRINT "Corresponding Time:        t = "; INT((tc) + .5)
LOCATE 8, 23: PRINT "Corresponding Speed:       v = "; INT((vc) + .5)
LOCATE 9, 23: PRINT "Corresponding Angle:     a="; INT((ac) * 1000 + .5) / 1000
LOCATE 10, 23: PRINT "Cross DEFLECTION:   z="; INT((cw * (tc - xc / vo1)) + .5)
LOCATE 12, 23: PRINT "AT TIME            T : "; INT((tt) + .5)
LOCATE 13, 23: PRINT "Parachutist Location   : "; "("; INT((xt) + .5); ",";
    INT((yt) + .5); ")"
LOCATE 14, 23: PRINT "Corresponding Speed : "; INT((vt) + .5)
LOCATE 15, 23: PRINT "Corresponding Angle : "; INT((at) * 1000 + .5) / 1000
LOCATE 16, 23: PRINT "Cross DEFLECTION   : "; INT((cw * (tt - xt / vo1)) + .5)
LOCATE 18, 23: PRINT "Drag Coefficient         : "; koef
ELSE
GOTO f:
END IF
END


SUB c (koef, a)
koef = koef
END SUB


SUB InfHyres (x0, y0, z0, v0, w0, a, koef, xc1, h0, vo, mas, xx1, alt, time, cd,
    Wind, cw, vo1, ta, ta1, pa, ea, pa1)
LOCATE 5, 13: INPUT "Launching Altitude [m]        = "; v0
```

```
LOCATE 6, 13: INPUT "Altitude of Parachute Deployment = "; xx1
LOCATE 7, 13: INPUT "Launching Speed [m/s]        = "; y0
LOCATE 8, 13: INPUT "Launching Angle [Degree]      = "; z0
LOCATE 9, 13: INPUT "Range Wind Speed           = "; Wind
LOCATE 10, 13: INPUT "Cross Wind Speed          = "; cw
LOCATE 11, 13: INPUT "Time of Interest          = "; time
LOCATE 12, 13: INPUT "Coefficient b           = "; cd
LOCATE 13, 13: INPUT "Mass of Parachutist         = "; mas
LOCATE 14, 13: INPUT "Temperatur of Air [Celsius] = "; ta
LOCATE 15, 13: INPUT "Atmospheric Pressure [mm]  = "; pa
LOCATE 16, 13: INPUT "Pressure of Air Vapor [mm]  = "; ea
LOCATE 17, 13: INPUT "Integration Step: 0.01       = "; h0
ta = ta + 273.15:  pa1 = ta / (1 - .3785 * ea / pa): ta1 = (289.08 / pa1)


vo = v0: alt = v0: vol = y0: a = z0: xc1 = 0: koef = cd / mas
y0 = SQR(vol ^ 2 + Wind ^ 2 - 2 * vol * Wind * COS(a * 3.141516954# / 180))
y0 = SQR(y0 ^ 2 + cw ^ 2)
y0 = y0 * COS(a * 3.141516954# / 180)
z0 = TAN(z0 * 3.141516954# / 180)
z0 = z0 / (1 - Wind / (vol * COS(a * 3.141516954# / 180)))
CLS
END SUB


SUB menu (cog, cof, xf, yf, xfu, yfu, t$)
COLOR cog, cof
LOCATE xf - 1, yf: PRINT t$
LOCATE xf, yf: PRINT "É" + STRING$(yfu - yf, 205) + "»";
FOR i = xf + 1 TO xfu
LOCATE i, yf: PRINT "º" + SPACE$(yfu - yf) + "º";
NEXT
LOCATE xfu + 1, yf: PRINT "È" + STRING$(yfu - yf, 205) + "¼";
END SUB


SUB NPkoef (k, L, r, q, h, y1, z1, v1, w1)
k = h * y1: L = h * z1
r = h * v1: q = h * w1
END SUB
```

```
SUB NPxyzvw (nk, x, x0, y, y0, z, z0, v, v0, w, w0, h, h0, k, L, r, q)
IF nk = 1 THEN
x = x0: y = y0: z = z0
v = v0: w = w0: h = h0
GOTO fund:
END IF

IF nk = 2 OR nk = 3 THEN
x = x0 + (.5 * h): y = y0 + (.5 * k)
z = z0 + (.5 * L): v = v0 + (.5 * r)
w = w0 + (.5 * q)
GOTO fund:
END IF

IF nk = 4 THEN
x = x0 + h: y = y0 + k: z = z0 + L
v = v0 + r: w = w0 + q
END IF
fund:
END SUB

SUB y1z1v1w1 (x, y, z, v, w, y1, z1, v1, w1, koef, ta1, pa)
yy = y * SQR(1 + z ^ 2)
y1 = -1*koef*(pa/750)*((289.08 - .006328*(v)) / 289.08)^4.4*(ta1*yy)^2 / (yy)
z1 = -9.80665 / y ^ 2
v1 = z
w1 = 1 / y
END SUB
```

REFERENCES

1. Hayden, R., Almgren, T., and Thomas, K., McDonald, W. T., *Sierra's Exterior Ballistics*, 5th ed., http://www.exteriorballistics. com/ebexplained/index.cfm.
2. Hurley, J. P., and Garrod, C., *Principi Di Fisica*, Zanichelli, 1986.
3. Klimi, G., *Exterior Ballistics with Applications—Skydiving, Parachute Fall, Flying Fragments*, Xlibris, 2008.
4. McCoy, R. L., *Modern Exterior Ballistics*, Schiffer Publishing Ltd., 1999.
5. Mori, E., *Balistica teorica e pratica*, http://www.earmi.it/balistica"
6. Mucinov, S.S., Shevcenko, N.A., *Zadacnik po Osnovami Strelbi is Strelkovogo Oruzie*, Moscow, 1964.
7. Okunev, B. H, *Fundamentals of Ballistics*, Vol.1, Book 2, Moscow, 1943.
8. Rinker, R. A., *Understanding Firearm Ballistics*, Mulberry House Publishing, 6th Ed, 2005.
9. Schaefer, J. C., *A Short Course in External Ballistics*, http://www. frfrogspad.com.
10. Shapiro, J. M., *Vneshnaja Balistika*, Oborongiz 50'.

Errata
For the
Exterior Ballistics with Applications
Skydiving, Parachute Fall, Flying Fragments,
by Gjergj Klimi,
Xlibris Corporation, 2008

Note: To locate the error (word, formula or equation) we show the page (P), and the line number.

For example, **P13, line 16** means that the error is in page 18, line 7 from above (the line is counted from the first row).

P. 20, line 2B, means that the error is in page 20, line 2 from bottom (the line is counted from the foot of the page.

A **formula or a function** in any page is counted as one line, while the **figure** is not counted as a line.

P.13, line 16, is written:	19	correct:	20
P.18, line 7, is written:	air	correct:	sound
P. 20, line 2B, is written:	4	correct:	5
P. 26, line 6B, is written:	g=9.0665 m/s	correct:	g=9.80665m/s
P. 27, line 14, is written:	(3)	correct:	(1.3)
P.28, line 5B, is written:	(4)	correct:	(1.2.4)
P28, line 3B, is written:	(4)	correct:	(1.2.4)
P.29, line 7B, is written:	(1.2.3)	correct:	(1.3.3)
P.30, line 10, is written:	(1.3.4)	correct:	(1.2.4)
P. 37, line 7B, is written:	(4)	correct:	(1.7.4)
P. 38, line 1, is written:	(2)	correct:	(1.7.2)
P.38, line 4, is written:	(11)	correct:	(1.7.11)
P.38, line 7, is written:	(7)	correct:	(1.7.7)
P.49, line 7-8, is written:	axis of the muzzle (firing mechanisms)	correct:	line of departure
P. 51, line 3, delete:	(aiming)		

P. 55, line 6, is written: x   correct: x_T

P. 59, line 13-14, is written: , for the same Mach   correct: are respectively
                                number, are

P. 65, line 12, and line 14   4.7302   correct: 4.732
(formula), is written:

P. 65, line 6B, delete: standard

P. 65, line 5B, is written: function   correct: functions

P.66, line, 5B (footnote), is   has   correct: was
written:

P.72, line 9, is written: perform   correct: performed

P.77, line 4, is written: m above sea   correct: meters above
                                                                          the sea

P.79-P.80, in all formulas   1.42233433   correct: 1.4223
(2.4.4), (2.4.5)(2.4.7),
(2.4.9) and (2.4.10) is
written:

P. 82, line 6 (formula), is   1.42233433   correct: 1.4223
written:

P.88, line 11, is written: ballistics   correct: drag

P. 88, line 7B-6B, delete:

            Substituting for $f_D(v)$ the approximate Siacci function
            $K_D(v) = (v - 240)/3$ and then solving
Write: Solving

P.88, line 4B (formula), is written:

$$C_D(\overline{v}/a) = f_D(\overline{v}) \div [4.732 \cdot 10^{-4} \overline{v}^2 c_D(\overline{v}/a)]$$

correct: $C_D(\overline{v}/a) = f_D(\overline{v}) \div (4.732 \cdot 10^{-4} \overline{v}^2)$

P. 90, formula (2.6.20), is written:

$$K_D(v) = \begin{cases} 1.212 \cdot 10^{-4} v^2 & for \quad v \le 256/v_0 \\ (v-240)/3 & for \quad v > 256/v_0 \end{cases}$$

Correct: $K_D(v) = \begin{cases} 1.212 \cdot 10^{-4} v^2 & for \quad v \le 256 \\ (v-240)/3 & for \quad v > 256 \end{cases}$

P.91, line 3B, delete: equation (2.6.17) and (2.6.18), using (2.6.20)

P. 98, line5B, delete: (1)

P. 107, line 6B (formula), is written: 1.42233433 correct: 1.4223

P. 109, formula (2.9.6), is written:

$$K_D(v) = \begin{cases} 1.212 \cdot 10^{-4} v^2 & for \quad v \le 256/v_0 \\ (v-240)/3 & for \quad v > 256/v_0 \end{cases}$$

Correct: $K_D(v) = \begin{cases} 1.212 \cdot 10^{-4} v^2 & for \quad v \le 256 \\ (v-240)/3 & for \quad v > 256 \end{cases}$

P. 110, formula (2.9.10), is written: 1.42233433 correct: 1.4223

P. 112, formula (2.10.13, is written:

$$K_D(v) = \begin{cases} 1.212 \cdot 10^{-4} v^2 & for \quad v \le 256/v_0 \\ (v-240)/3 & for \quad v > 256/v_0 \end{cases}$$

Correct: $K_D(v) = \begin{cases} 1.212 \cdot 10^{-4} v^2 & for \quad v \le 256 \\ (v-240)/3 & for \quad v > 256 \end{cases}$

P. 118, line 4B (formula), is written: $K_D(v) = 1.212 \cdot 10^{-4} v_0^2$
Correct: $K_D(v) = 1.212 \cdot 10^{-4} v^2$

P. 119, line 2 (formula), is written: $b = 1.212 \cdot 10^{-4} c \cdot h(y) \cdot v_0^2 \cdot$
Correct: $b = 1.212 \cdot 10^{-4} c \cdot h(y),$

P. 119, formula (2.11.12), is written: $v_m = (-\frac{g v_0^2}{b} \sin \alpha_m)^{1/2}$

Correct: $v_m = (-\frac{g}{b} \sin \alpha_m)^{1/2}$

P. 121, p. 14, is written: 289.06 correct: 289.08

P. 162, formula (3.5.2),

is written: $\bar{y}=\dfrac{g}{2\cdot u_0^2}\cdot\bar{x}^2+\dfrac{gB(u_0-240)}{3u_0^4}\bar{x}^3+gB^2(u_0-240)\dfrac{u_0-320}{4v_0^6}\bar{x}^4.$

correct: $\bar{y}=\dfrac{g}{2u_0^2}\bar{x}^2+\dfrac{gB(u_0-240)}{3u_0^4}\bar{x}^3+gB^2(u_0-240)\dfrac{u_0-320}{4u_0^6}$

P. 165, line 2 (in the formula), is written: $\dfrac{u_0-320}{4v_0^6}$ correct: $\dfrac{u_0-320}{4u_0^6}$

P. 165, line 10B (in the formula), is written: $\dfrac{u_0-320}{4v_0^6}$ correct: $\dfrac{u_0-320}{4u_0^6}$

P.167, line 2 (in the formula), is written: $\dfrac{u_0-320}{4v_0^6}$ correct: $\dfrac{u_0-320}{4u_0^6}$

P. 167, line 5B, (in the formula), is written: $\dfrac{u_0-320}{4v_0^6}$ correct: $\dfrac{u_0-320}{4u_0^6}$

P. 225, formula (5.2.2)

is written:

$$\begin{cases} \dfrac{dp}{du}=\dfrac{1}{B(u-240)\cdot u\cdot\cos^2\alpha_0} \\[2mm] \dfrac{dt}{du}=-\dfrac{g}{B(u-240)\cdot\cos\alpha_0} \\[2mm] \dfrac{dx}{du}=-\dfrac{gu}{B(u-240)} \\[2mm] \dfrac{dy}{du}=-p\dfrac{gu}{B(u-240)} \end{cases}$$

correct:
$$\begin{cases} \dfrac{dp}{du} = \dfrac{1}{B(u-787.4)\cdot u\cdot \cos^2\alpha_0} \\[2mm] \dfrac{dt}{du} = -\dfrac{1}{Bg(u-787.4)\cdot \cos\alpha_0} \\[2mm] \dfrac{dx}{du} = -\dfrac{u}{Bg(u-787.4)} \\[2mm] \dfrac{dy}{du} = -p\dfrac{u}{Bg(u-787.4)} \end{cases}$$

P. 227, line6B, delete: the first and

P. 227, line6B, is written: ones correct: one

P. 229-P 230, delete all the following:

(a) For antiaircraft fire, substituting in the above equation $p_0 = \tan\alpha_0$, $B = \beta \cdot b$ and $\beta = h(\bar{y})/\cos\alpha_0$ we obtain

$$\tan\alpha_0 = -\frac{1}{240h(\bar{y})b\cos(\alpha_0)}\cdot[\frac{A(u)-A(v_0)}{D(u)-D(v_0)} - J(v_0)] \qquad (5.2.25)$$

Multiplying both sides of (25) by $\cos(\alpha_0)$ we find the following formula to estimate the launching angle as a function of pseudo-speed at the point of impact:

$$\sin(\alpha_0) = -\frac{1}{240h(\bar{y})b}\cdot[\frac{A(u)-A(v_0)}{D(u)-D(v_0)} - J(v_0)] \qquad (5.2.26.a)$$

P. 230, line 2, delete: (b)

P. 230, line 5, is written: (5.2.26a) correct: (5.2.26)

P. 243, line 10B, is written: example 2.8.2 correct: example 3, section 2.6

P. 243, line 8B, is written: example 2, section 2.8
correct: example 3, section 2.6

P. 302, line 14B, is written: "the correct:"

P. 307, line 7, is written:

$$\frac{\partial x}{\partial T_0} = (1 - \frac{v_0}{2x}\frac{\partial x}{\partial v_0})\frac{x}{T_0} = (1 - \frac{(735)}{2(300)}\cdot(0.688347))\frac{300}{288.15} = 0.15677$$

Correct:

$$\frac{\partial x}{\partial T_0} = (1 - \frac{v_0}{2x}\frac{\partial x}{\partial v_0})\frac{x}{T_0} = (1 - \frac{(735)}{2(300)}\cdot(0.688347))\frac{300}{299.08} = 0.15104$$

P. 307, line 7B, is written: $\Delta x_2 = \frac{\partial x}{\partial T_0}dT_0 = (0.15677)(10) = 1.5677m$

Correct: $\Delta x_2 = \frac{\partial x}{\partial T_0}dT_0 = (0.15104)(10) = 1.51m$

P. 398, line 3B, is written: $\tau_{0N} = 289.06°K$ correct: $\tau_{0N} = 289.08°K$

P. 405, formula (8.5.15), is written: $H(y) = (\frac{\tau}{\tau_0})h_\tau(y)$

correct: $H(y) = (\frac{\tau_{0N}}{\tau_0})h_\tau(y)$

P. 405, formula (8.5.16), is written: $h_\tau(y) = (\frac{\tau_0}{\tau})H(y)$

correct: $h_\tau(y) = (\frac{\tau_0}{\tau_{0N}})H(y)$

P. 405, line 6, is written: dH in $H(y)$
correct: $dH = p - p_{0N}$ in $H(y) = p / p_{0N}$,

P. 405, line 6, is written: dh correct: $dh = \rho - \rho_{0N}$

P. 405, formula (8.5.17), is written: $dh = (\frac{\tau_0}{\tau})dH$ correct: $dh=(\frac{\tau_0}{\tau_{0N}})dH$

P. 405, line 6B, is written: (18) correct: (8.5.18)

P. 405, formula (8.5.19), is written: $h' = h_\tau(y)(1+\frac{dH}{H(y)}) = h_\tau(y)\cdot(1+\frac{dp}{p})$

Correct:

$$h'=h_\tau(y)(1+\frac{p_0-p_{0N}}{p_{0N}})=h_\tau(y)\cdot\frac{p_o}{p_{0N}}$$

P. 408, formula (8.5.27), is written:

$$\begin{cases} \dfrac{dv_x}{dx} = -ch_\tau(y)\cdot\dfrac{\tau_0}{\tau_{0N}}\dfrac{V(\tau_{0N}/\tau_0)^{1/2}-240}{3v} \\[2mm] \dfrac{dp}{dx}=-\dfrac{g}{v_x^2} \\[2mm] \dfrac{dt}{dx}=\dfrac{1}{v_x} \\[2mm] \dfrac{dy}{dx}=p \end{cases}$$

Correct:

$$\begin{cases} \dfrac{dv_x}{dx} = -ch_\tau(y)\cdot\dfrac{v(\tau_{0N}/\tau_0)^{1/2}-240}{3v} \\[2mm] \dfrac{dp}{dx}=-\dfrac{g}{v_x^2} \\[2mm] \dfrac{dt}{dx}=\dfrac{1}{v_x} \\[2mm] \dfrac{dy}{dx}=p \end{cases}$$

P. 408, formula (8.5.28), is written

$$\begin{cases} \dfrac{dv_x}{dx} = -1.212 \cdot 10^{-4} ch_\tau(y) \cdot \dfrac{V^2}{v} \\[2mm] \qquad \dfrac{dp}{dx} = -\dfrac{g}{v_x^2} \\[2mm] \qquad \dfrac{dt}{dx} = \dfrac{1}{v_x} \\[2mm] \qquad \dfrac{dy}{dx} = p \end{cases}$$

Correct: $$\begin{cases} \dfrac{dv_x}{dx} = -1.212 \cdot 10^{-4} ch_\tau(y) \cdot \dfrac{(v \cdot \tau / \tau_{ON})^2}{v} \\[2mm] \qquad \dfrac{dp}{dx} = -\dfrac{g}{v_x^2} \\[2mm] \qquad \dfrac{dt}{dx} = \dfrac{1}{v_x} \\[2mm] \qquad \dfrac{dy}{dx} = p \end{cases}$$

P. 408, formula (8.5.29), is written:
$$V = [(\vec{v} - \vec{w}_x)^2]^{1/2} = (v^2 + w_x^2 - 2v \cdot w_x \cdot \cos\alpha)^{1/2}$$

Correct: $v_0 = [(\vec{v}'_0 - \vec{w}_x)^2]^{1/2} = (v_0'^2 + w_x^2 - 2v'_0 \cdot w_x \cdot \cos\alpha_0)^{1/2}$,
is the relative initial speed of the projectile while v'_0 is the speed of the projectile that considers the other changes; see formula (8.6.8).

P. 409, formula (8.5.32), is written: $h' = h_\tau(y) \cdot (1 + \dfrac{dp}{p})$

Correct: $h' = h_\tau(y) \dfrac{p}{p_{ON}}$

P. 410, line 2B (do not count footnote lines), is written: uses correct: use

P. 412, formula (8.6.1), is written:

$$\left\{ \begin{array}{l} \dfrac{dv_x}{dx} = -c'h_\tau(y)(1+\dfrac{dH}{H}) \cdot \dfrac{\tau_0}{\tau_{ON}} \dfrac{V(\tau_{ON}/\tau_0)^{1/2} - 240}{3v} \\[2ex] \dfrac{dp}{dx} = -\dfrac{g}{v_x^2} \\[2ex] \dfrac{dt}{dx} = \dfrac{1}{v_x} \\[2ex] \dfrac{dy}{dx} = p \end{array} \right.$$

Correct:
$$\left\{ \begin{array}{l} \dfrac{dv_x}{dx} = -c'h_\tau(y) \cdot \dfrac{p_0}{p_{ON}} \cdot \dfrac{v(\tau_{ON}/\tau_0)^{1/2} - 240}{3v} \\[2ex] \dfrac{dp}{dx} = -\dfrac{g}{v_x^2} \\[2ex] \dfrac{dt}{dx} = \dfrac{1}{v_x} \\[2ex] \dfrac{dy}{dx} = p \end{array} \right.$$

P. 412, formula (8.6.2), is written:

$$\left\{ \begin{array}{l} \dfrac{dv_x}{dx} = -1.212 \cdot 10^{-4} c'h_\tau(y)(1+\dfrac{dH}{H}) \cdot V^2 \dfrac{1}{v} \\[2ex] \dfrac{dp}{dx} = -\dfrac{g}{v_x^2} \\[2ex] \dfrac{dt}{dx} = \dfrac{1}{v_x} \\[2ex] \dfrac{dy}{dx} = p \end{array} \right.$$

Correct:

$$\begin{cases} \dfrac{dv_x}{dx} = -1.212 \cdot 10^{-4}\, c'h_\tau(y)\, \dfrac{p_0}{p_{0N}} \cdot \dfrac{(v \cdot \tau_{0N}/\tau)^2}{v} \\[2mm] \dfrac{dp}{dx} = -\dfrac{g}{v_x^2} \\[2mm] \dfrac{dt}{dx} = \dfrac{1}{v_x} \\[2mm] \dfrac{dy}{dx} = p \end{cases}$$

P. 412, formula (8.6.4), is written: $h_\tau(y) = \dfrac{\rho}{\rho_0} = \left(\dfrac{\tau_0 - 0.006328y}{\tau_0}\right)^{4.4}$

correct: $h_\tau(y) = \dfrac{\rho}{\rho_{0N}} = \left(\dfrac{289.08 - 0.006328y}{289.08}\right)^{4.4}$

P. 412, formula (8.6.6) is written: $V = (v^2 + w^2 - 2v \cdot w \cdot \cos\alpha)^{1/2}$

Correct: $v_0 = (v_0'^2 + w^2 - 2v_0 \cdot w \cdot \cos\alpha_0)^{1/2}$

is the relative initial speed of the projectile while v_0' is the speed of the projectile that considers the other changes; see formula (8.6.8).

P. 414, line 4B (footnote); is written:	Artillery is	correct:	, is artillery
P. 491, line 7, is written:	793.92	correct:	895m
P. 491, line 8, is written:	67.46	correct:	66m/s
P. 491, line 8, is written:	86.20	correct:	87.5 degree
P. 491, line 10, is written:	518, 2255	correct:	512, 2269
P. 491, line 11, is written:	66.60	correct:	58m/s
P. 491, line 11, is written:	-52.96	correct:	-57.77
P. 496, line 9B, is written:	2311	correct:	2319

P. 496, line 8B, is written: 65.5; Falling Angle = -47.40
correct: 57; Falling Angle = -52.77

P. 496, line 5 (formula), is written: 13.44   correct: 14

P. 510, line 6, is written: 589; y-coordinate of Parachutist = 1987
Correct: 599; y-coordinate of parachutist, 1999

P. 510, line 7, is written: Speed of Parachutist = 67.89; Falling Angle = -61.44
Speed of Parachutist, 63; Falling Angle, 65.76

P. 511, line 6, is written: $x = 589 + 30 = 619m$, $y = 1987 - 110 = 1877m$.

Correct: $x = 599 + 30 = 629m$, $y = 1999 - 110 = 1889m$.

P. 511, line 5B, is written:

$$v_e = (\frac{mg}{b \cdot h(y)})^{1/2} = 5.263 \cdot h(y)^{-1/2} = 5.263 \cdot (\frac{289.06}{289.06 - 0.006328 \cdot y})^{0.5}$$

Correct:

$$v_e = (\frac{mg}{b \cdot h(y)})^{1/2} = 5.263 \cdot h(y)^{-1/2} = 5.263 \cdot (\frac{289.08}{289.08 - 0.006328 \cdot y})^{0.5}$$

P. 512, line 10 (formula), is written: $x = 619 + 3410 = 4,029mm$

Correct: $x = 629 + 3410 = 4,039mm$

P. 554, line 5 and line 6, delete all the following:

(a) For Up-hill Fire

$$\sin(\alpha_0) = -\frac{1}{240h(\bar{y})b} \cdot [\frac{A(u) - A(v_0)}{D(u) - D(v_0)} - J(v_0)]$$

P. 554, line 7, delete: a.

P. 587, line 7, is written: is   correct: are

P. 587, line 11, is written: Nonstandard   correct: Standard

P. 587, line 1B, is written: is   correct: are